Planted Forests

Uses, Impacts and Sustainability

Planted Forests

Uses, Impacts and Sustainability

Edited by

Julian Evans

Published by
Food and Agriculture Organisation of the United Nations

and

www.cabi.org

CABI is a trading name of CAB International

CABI Head Office
Nosworthy Way
Wallingford
Oxfordshire OX10 8DE
UK

Tel: +44 (0)1491 832111
Fax: +44 (0)1491 833508
E-mail: cabi@cabi.org
Website: www.cabi.org

CABI North American Office
875 Massachusetts Avenue
7th Floor
Cambridge, MA 02139
USA

Tel: +1 617 395 4056
Fax: +1 617 354 6875
E-mail: cabi-nao@cabi.org

A catalogue record for this book is available from the British Library, London, UK.

Library of Congress Cataloging-in-Publication Data
Planted forests : uses, impacts, and sustainability / edited by Julian Evans.
p. cm.
Includes bibliographical references and index.
ISBN 978-1-84593-564-1 (alk. paper)
1. Afforestation—Social aspects. 2. Afforestation—Environmental aspects.
3. Tree farms—Social aspects. 4. Tree farms—Environmental aspects.
5. Sustainable forestry. I. Evans, Julian, 1946-

SD409.P685 2009
333.75'152–dc22

2009018693

Published jointly by CAB International and FAO.
Food and Agriculture Organization of the United Nations (FAO)
Viale delle Terme di Caracalla, 00153 Rome, Italy
website: www.fao.org

ISBN: 978 1 84593 564 1 (CABI)
ISBN: 978 92 5 106222 7 (FAO)

Typeset by AMA DataSet Ltd, Preston, UK.
Printed and bound in the UK by MPG Books Group.

The paper used for the text pages in this book is FSC certified.
The FSC (Forest Stewardship Council) is an international network to promote responsible management of the world's forests.

Contents

Contributors

J. Evans, Consultant, formerly Professor of Forestry, Imperial College London, and Chief Research Officer(S), British Forestry Commission.

J. B. Carle, Chief, Forest Resources Development Service, Food and Agriculture Organization (FAO), Rome, Italy.

J. B. Ball, Consultant, Forest Resources Development Service, Food and Agriculture Organization (FAO), Rome, Italy.

A. Del Lungo, Forestry Officer, Forest Resources Development Service, Food and Agriculture Organization (FAO), Rome, Italy.

D. A. Neilson, Managing Director, D. A. Neilson and Associates, Rotorua, New Zealand.

L. P. B. Holmgren, Director, Environment, Climate Change and Bioenergy Division, Food and Agriculture Organization (FAO), Rome, Italy.

Foreword

The goal of sustainable forest management has received considerable attention in international negotiations. The Rio Declaration (UNCED) and several of the United Nations conventions, as well as the United Nations Forum on Forests (UNFF) and other international processes, meetings and key publications, have recognized the critical role of forestry, including planted forests, in achieving sustainable development and mitigating the effects of climate change.

Planted forests have been a legitimate land use for centuries. They have expanded rapidly in both area and impacts in recent decades. In 2005 it was estimated that planted forests constituted only 2% of global land area (7% of forest area), or about 271 million hectares. The potential industrial roundwood production from planted forests in 2005 was estimated at 1.2 billion m^3 or about two-thirds of global industrial roundwood needs. The significance of planted forests and recognition of their contributions to a range of development goals are anticipated to increase in coming decades. Planted forests provide not only wood, fibre and fuel, but also other non-wood forest products. Moreover, they sequester carbon, rehabilitate degraded lands, help in restoring landscapes, protect watersheds and agricultural soils, and provide recreational areas and amenities. There is increasing public awareness that wood products have advantages over competing products made of other materials (cement, plastics and metal) in that wood is renewable, energy efficient and environmentally friendly if managed in a responsible manner. Intensively managed planted forests are an effective land use for these purposes.

Ownership of planted forests globally, calculated on an area basis, is: government, 50%; smallholders, 32%; and private-sector corporate, 18%. The corporate private sector employs forestry professionals who deploy improved genetic stock and nursery practices, apply intensive silvicultural management and invest in fire and forest health protection that result in high productivity and quality forest products that command premium prices on the market place. The application of new knowledge and technology in planning and improved practices is not always adopted in government and smallholder plantings, particularly in developing countries.

A lack of knowledge, capacity and capability in providing enabling policies, laws, regulations, plans and technical support systems, particularly in developing countries, have led to some planted forest investments causing land-use, social and environmental conflicts, as well as resulting in poor forest health, productivity and returns on investment. Through a multi-stakeholder process, FAO prepared *Responsible Management of Planted Forests: Voluntary Guidelines* (FAO, 2006b) and has embarked upon a programme of country capacity-building to balance the social, cultural, environmental and economic dimensions of planted forest in landscape management approaches to increase the contribution of planted forests to sustainable livelihoods and land use.

Policy makers, managers and forest investors must consider the unique context in which they are investing in planted forests and respond to the prevailing and perceived driving forces, including socio-economic conditions, markets, consumer demand and new technologies. In each context they must consider the production technologies, market place, the wood industries sectors and also social demands and environmental covenants.

FAO is committed to strengthening country capacity in formulating enabling policies and technical standards for responsible management of planted forests. The goal is to increase their provision of goods and services towards achievement of sustainable livelihoods and land use and, in particular, to mitigate the effects of climate change and provide a renewable source of wood, fibre and fuel.

Jan Heino
Assistant Director-General
Forestry Department
Food and Agriculture Organization of the United Nations
Italy

Preface

This book provides a synthesis of the uses, impacts and sustainability of planted forests by looking at the past, outlining the present situation and highlights the outlook and issues for the future. The principles and key considerations of the *Voluntary Guidelines for Responsible Management of Planted Forests* (FAO, 2006b) are cross-cutting throughout the book.

The first introductory chapter sets the scene for the book, briefly introducing the role planted forests play, their strengths and weaknesses, and their potential for the future.

Chapter 2 gives the origins of early planting and the evolution of planted forests in recent history, awareness of the impacts of sound silviculture and more recently the need to meet social, cultural, environmental and economic objectives and provide a wide range of goods and services.

Chapter 3 details, and gives some history to, the issue of definition of planted forests highlighting the continuum of different forests and intensities of management (including planted forests) and trees outside forests across the landscape. The chapter introduces the management objectives, whether productive or protective.

Chapter 4 synthesizes highlights of the FAO's *Global Planted Forests Thematic Study* (Del Lungo *et al*., 2006), including the results and analysis of the global survey of planted forest 2005. Survey results are summarized on an area basis (1990, 2000, 2005), according to forest plantations, planted semi-natural forests and total planted forests, by productive or protective purpose. Additional information according to ownership, species, growth rates, age classes, rotation and end uses is also summarized. Detailed area tables by country are available in the Appendix.

Chapter 5 summarizes the key findings of FAO's *Global Planted Forests Outlook 2005-2030* (Carle and Holmgren, 2007), which highlights that although planted forests cover less than 3% of land area, they contribute a considerably higher proportion of overall goods (wood, fibre, fuel) and environmental and social services, now, and increasingly in the future.

Chapter 6 outlines the different roles of planted forests including social, environmental, and ecological roles in different contexts. The multiple facets of planted forests are highlighted for production of wood, fibre, fuel; soil and water protection, mitigating the impacts of climate change (carbon sinks or carbon sequestration); and amenity, recreational or landscape restoration.

Chapter 7 focuses upon the critical aspects of policy, institutional and ownership of planted forests. The different perspectives of private sector corporate and smallholder are highlighted from an investment perspective. The chapter draws upon a study commissioned by FAO *Corporate Private Sector Investment Dimensions of Planted Forest Investments* (Neilson, 2007a).

Chapter 8 reviews the issues relating to sustainability of planted forests through subsequent rotations. Issues such as planted forests and their management impacts on soils, nutrient balance, threats (insects, diseases and other pests) and site changes are raised as well as risks they pose such as that of invasive species. Management interventions to minimize risks are suggested.

Chapter 9 summarises the key issues drawn from each chapter and concludes that planted forests fulfil a critical role in the social, environmental and economic dimensions of sustainable forest management and these will increase in the future.

Owing to the significance emerging from the data about the importance of planted forests worldwide, FAO decided to publish this synthesis of its working papers both to draw out crucial impacts and issues and, by inviting an external editor, to set planted forests in a wider context. We hope the book will be of assistance to many from policy maker to practitioner.

Acknowledgements

Sincere thanks are extended to all those who contributed to the organization and contents of this book. Several people were involved in addition to the authors named.

Supervision of this book was provided by Jim Carle, Chief, Forest Resources Development Service (FOMR), FAO, who provided the resources and guidance on scope and technical content.

Katrine Myrseth and Kate Beauchamp, Imperial College London, and Eleanor Harland, Forest Research, provided assistance with background research and help with the bibliography. Martin Evans of Agriprojects Consulting Ltd, Cambridge, provided input on the fast-developing biofuels industry. Hugh Evans, Forest Research UK, summarized pest and disease issues affecting planted forests worldwide.

Permission is given by John Turnbull, co-author of *Plantation Forestry in the Tropics* (2004, 3rd edition), and by Oxford University Press, to reproduce material from that book and which appears here mainly in Chapter 2.

Special thanks to Alberto Del Lungo, FOMR, FAO, who provided technical support in the planted forest database, figures and photos. Graciela Andrade, FOMR, FAO, supported the project in countless ways. Rachel Tucker, Publishing Planning and Rights Manager, FAO, provided the critical liaison between CABI and FAO for development of the project into book form. Stephen Evans provided invaluable support to the editor.

Illustrations: all photographs were provided by the editor (J. Evans) or from FAO's Photo Library except fig. 8.9 which is reproduced by permission of Professor J.L. Innes, University of British Columbia.

Fig. 6.7 is reproduced in its present form from Evans, J. and Turnbull, J. (2004) *Plantation Forestry in the Tropics*, 3rd edition, Oxford University Press, Oxford (Fig. 22.1, page 354).

Fig. 8.4 is reproduced in its present form from Evans, J. (1999) *The Sustainability of Forest Plantations – the Evidence*. Department for International Development, London (page 18). The original was redrawn with permission of Dr E.K.S. Nambiar.

1 Introduction

J. EVANS

1.1 Planted Forests at a Critical Point

From time to time there comes a tipping point when an occasion or set of circumstances clearly set the agenda for the future. Such a critical point may have been reached with the ascendancy of planted over natural forests for supplying many industrial commodities, some environmental services and even a few wildlife benefits. This ascendancy can be traced in FAO's own work, and particularly its publications, from its inception nearly 60 years ago through the seminal conference on Man-Made Forests and their Industrial Importance in 1967 and numerous technical reports, newsletters and successive World Forestry Congress proceedings, to this book today. Planted forests have come of age.

In asserting this view – that planted forests are playing a role far in excess of what their actual area might suggest – limitations are recognized. They do not and cannot substitute natural forest formations: they are not an alternative but are complementary. They may help ease some pressures on natural forests, but that is unlikely to be their major role, desirable as this might be. Planted forests, in all their variety, offer major opportunities but are no panacea to the ills that beset the world's forests at large, namely, destruction and deforestation, the loss of ecosystems and environmental services, and perhaps most important of all, the loss of somewhere to live for many groups of people too often on the fringe of society. Tree planting and planted forests have a role to play and are part of the solution to these ills.

Identifying the role planted forests play, examining their strengths and weaknesses, and exploring in what ways they are part of the future of the world's forest resources is the aim of this book. Planted forests are often no more than the foresters' equivalent of a farm crop, but which through sound management can deliver some benefits beyond those purely of production. It is here that the farm analogy breaks down. First, as stands of trees grow, mature and regenerate, they can certainly continue to be worked as 'crops', but many are managed in

Fig. 1.1. Planted *Fagus sylvatica* in southern England exhibiting features of natural woodland.

ways through which they come to appear less and less artificial and take on features of naturalness (Fig. 1.1). The latter has commonly occurred when native species constitute the planted forest. Second, all planted forests can offer several, even many, benefits over and above that of production, but none ever truly substitutes for natural forest. This book recognizes both these opportunities and the limitations. The outlook is neither infinitely rosy nor doom laden for planted forests. The attempt is made to bring a balanced view of a silvicultural tool that has come of age to serve some of humankind's needs.

The book draws on the latest research, both commissioned by FAO and in the published domain, and the experience of many people. But a 'health warning' is added. Inferences drawn rely on the reliability of the data and statistics available and their quality. In the chapters that follow this limitation is recognized. The point is made because it is a re-evaluation of data about planted forests that largely occasions this book.

1.2 Classifying Forest as 'Planted'

Classifying forests is surprisingly inexact for those that have been planted and therefore data concerning their extent are uncertain. There are, of course, reasons other than classification for such misreporting. For tropical and subtropical countries, Evans and Turnbull (2004) note several causes.[1] To these can be added the pressure to report what a nation's policy or plan stipulates, or a regional manager demands, rather than what is actually achieved. However, the issue here is not so much the figures themselves and these sorts of imperfections, but whether the extent of planted forests is poorly estimated because of what is or is not classified as 'planted'. The uncertainty has arisen because many forests that were planted long ago do not look like 'plantations' with all the perceived connotations the word conveys (Fig. 1.1).

One note of caution is that, in much of what has been written, even in the recent past, the terms 'plantation', 'forest plantation' and 'planted forest' have been used interchangeably as virtual synonyms (e.g. Savill *et al.*, 1997; Anon., 1999, 2003; Boyle *et al.*, 1999; Evans and Turnbull, 2004). Here a clear distinction is developed. 'Planted forest' includes all of what is generally understood as 'plantations' or 'forest plantation', but also incorporates other forest types originating largely or wholly from tree planting. What this means and what types of forest are included in this wider embrace is the purpose of Chapters 2 and 3.

1.3 Outline of the Book

To describe fully what 'planted forest' means and the implications this term has, it is important to sketch the history of tree planting, particularly the early history which is often forgotten, to place in context the questions and issues faced today. This is the subject of Chapter 2. On this foundation, in Chapter 3, the matter of definitions is explored, followed by a description of the *Global Forest Resources Assessment 2005* (hereafter FRA 2005) classification (FAO, 2005a). Thus Chapters 2 and 3 set the scene for the new data concerning planted forests reported in FRA 2005 and emerging from FAO's thematic study (Chapter 4) and their implications for forest production (Chapter 5). Chapter 6 analyses issues and explores in more depth the roles planted forests can be expected to play. Chapter 7 addresses the critical questions of who owns the world's planted forests, how this is changing, and what roles policy and institutional issues play. Chapter 8 examines the whole matter of sustainability of planted forests – how safe is it to rely on growing trees in this way. Conclusions are in Chapter 9. As well as the substantial bibliography, full data of planted forest areas for all the key countries are included in the Appendix.

1.4 Global Planted Forests Thematic Study

FAO's Global Planted Forest Thematic Study (FAO, 2006c) has been reported in working papers that detail the survey, responses, the associated desk study and results in status and trends in planted forests, as shown in Table 1.1. This book draws on all this research to present a detailed overview and draw conclusions.

The working papers themselves are aimed at a target audience of forestry planners and policy-makers as well as other interested parties, at national, regional and global levels, and are available from FAO.

Table 1.1. Key FAO working papers providing source material for this book.

No.	Title	Description
35	Global Planted Forests Thematic Study: Supplement to Forest Resources Assessment 2005 – Guidelines for National Reporting Tables for Planted Forests	Guidelines to national correspondents for completion of reporting tables in the planted forests survey
35a	Global Planted Forests Thematic Study: Country Responses to Reporting Tables for Planted Forests Survey	Country reporting tables for 36 countries with major planted forests areas
35b	Global Planted Forests Thematic Study: Supplementary Desk Study on Planted Forests	Supplementary desk study for 23 countries with significant areas of planted forests
38	Global Planted Forests Thematic Study: Results and Analysis	Main results, analysis, summary tables, figures, conclusions, recommendations and full data tables in annexes

Note

1 Areas reported 'should only be considered as approximate as the quality and quantity of plantation data are very dependent on the capacity of national forest inventory systems to collect and analyze data. Different sources often cite widely divergent figures for any one country. Some countries in official reports equate annual planting achievement with numbers of tree seedlings produced with no check on what land has actually been successfully planted; this invariably results in overestimates. In other cases the full ground area of poorly stocked plantations is also included which is misleading in terms of the quantity of standing timber' (Evans and Turnbull, 2004, p. 32).

2 The History of Tree Planting and Planted Forests

J. EVANS

2.1 Introduction

As noted in Chapter 1, classifying forests is surprisingly inexact for those that have been planted. One source of the uncertainty is many forests that were planted long ago do not look like 'plantations' today. To recall the history of tree planting helps focus on this issue of time. It is also a reminder that issues today were not of consequence in the past and vice versa, which further helps to bring perspective.

This outline of history draws in part – with permission as well as much modification – on two texts concerned with planted forests (Savill *et al.*, 1997; Evans and Turnbull, 2004). It also expands on several issues, particularly ones that bear on the question of definition and types of planted forests. The outline attempts to be global if, at best, no more than a sketch. In these ways the chapter sets the scene for those that follow.

2.2 Origins of Planting

People have been planting trees for thousands of years for food or other non-wood forest products, shelter, ornamental, ceremonial or religious purposes. The first woody species selected and planted, around 4000 BC, was probably the olive tree (*Olea europaea*) (Fig. 2.1), which has been cultivated in Greece at least since the Minoan era (3000 BC). The temple of Queen Hatshepsut, constructed in 1500 BC at Thebes, Egypt, has depictions of myrrh trees (*Commiphora myrrha*), introduced from Somalia, planted as sources of perfume, and Theophrastus reported trees of frankincense (*Boswellia* spp.) and myrrh planted on private estates in southern Arabia in the 4th century BC. There are also several biblical references to tree planting dating to 2000 BC or earlier, such as the Old Testament record of Abraham planting a tamarisk tree (*Tamarisk aphylla*) to commemorate the treaty of Beersheba (Gen. 21: 33).

Tree planting was practised in ancient times in Asia. The Chinese cultivated fruit trees, and grew pines for ornamental, religious and ceremonial purposes as long ago as 2000 BC (Valder, 1999). During the early part of the Chou Empire (*c.* 1100–256 BC) the Emperor established a forest service with the responsibility for preserving natural forest and reforesting denuded lands. The Han and Tang Dynasties (208 BC to AD 256) encouraged people to plant trees important for both food and timber production. In the Sung Dynasty (AD 420–589) direct planting of tree seeds for reforestation was widely practised and public land reforested by farmers became their property. Monographs were also prepared describing methods of planting and protection of the tung tree (*Aleurites* sp.), bamboos and other woody species (Wang, 1988). Ji Han's *Account of the plants and trees of the southern regions* is dated 304 AD. Chinese fir (*Cunninghamia lanceolata*), which today forms millions of hectares of planted forest in central and southern China, has a cultivation history of over 1000 years. In Korea, during the Shilla Kingdom, starting 57 BC, trees were planted around kings' tombs, in royal gardens, along roadsides, as shelterbelts, along rivers and on the coast for erosion control. In Sri Lanka, during the reign of the Sinhalese King Vijaya (*c.* 543 BC), village communities planted home gardens with flowering and fruit-bearing trees, and under King Dutugemunu (161–137 BC) forests were planted and rules made for forest protection and use of forest products (Winter, 1974). Probably the oldest planted tree still living, of which a reasonable historical records exists, is in the tropics, in Sri Lanka, where the Bodhi tree (*Ficus religiosa*) is recorded as being 'planted' miraculously at Anuradhapura in about 220 BC.

2.3 Planted Forests Before 1900

For this period it is convenient to distinguish between Europe, and by extension other temperate and 'Mediterranean type' regions, and the tropics and subtropics. Although there are many similarities in practices, what happened in Europe both influenced silviculture beyond its borders through the accidents of history – colonialism – and was massively influenced by what was found there. The renaissance sciences in the field of botany of plant systematics, classification and taxonomy, supremely associated with the Swede, Carolus Linnaeus, fathered the expeditions that scoured the globe for new species of trees and exotic plants to stock European arboreta – the zoos of botany – and the lands of the nobility. These collections were the bedrock of much of the present-day forest plantation resource.

2.3.1 Europe and related countries

Tree planting in Europe developed as a way of supplementing regeneration, changing the composition of forests, such as from beech (*Fagus sylvatica*) and firs (*Abies alba*) to spruce (*Picea abies*), and augmenting depleted forest cover (Koch and Skovsgaard, 1999). It began gradually in the Middle Ages. In Germany, sowing on a large scale took place as early as the 13th century in the forests

Fig. 2.1. Ancient olive trees in the Garden of Gethsemane, Jerusalem. Olive (*Olea europaea*) may be the earliest tree to have been brought into cultivation and planted.

around Nuremburg in order to re-forest the heavily exploited area (Ortloff, 1999). In the UK by the 1660s, John Evelyn urged in his great *Sylva* the repairing of the 'wooden walls' of England by tree planting. And, 80 years earlier, none other than Queen Elizabeth I had expressed concern over dwindling supplies of naval timber, which led to some renewed planting of oaks (*Quercus* spp.). Comparable exhortations and examples can be found for the same period in Germany and France. Colbert, and later Pannelier, were instrumental in the planting of several thousand hectares of mainly oak (*Quercus* spp.) forest in the Forêt de Compiegne near Paris in the 17th and 18th centuries. By the early 19th century most texts and treatises on forestry included many pages on tree planting and plantation establishment (Savill *et al.*, 1997).

Up until the early 19th century, nearly all planting was of native species in traditional forest areas: examples in the UK include the New Forest, the Chilterns and the Forest of Dean, and in France the Forêt de Compiegne (Fig. 2.2). Increasingly from this time tree planting also involved afforestation of bare land, such as *Pinus pinaster* on the Landes of France and spruce (*P. abies*) and pine (*P. sylvestris*) planting on old fields in the Vosges, or conversion of broadleaved forest to conifer. In Scotland, many larch (*Larix decidua*) plantations and beech (*F. sylvatica*) and pine (*P. sylvestris*) shelterbelts date back to this period. In Austria, as early as 1786 the Forest Ordinance recognized clear-cutting with artificial regeneration as the main silvicultural system in force (Troup, 1952). But it was in Germany that this new 'plantation forestry' developed most rapidly, largely through the influence of Cotta, such as the extensive spruce (*P. abies*) afforestation in Saxony.

Fig. 2.2. Much of the Forêt de Compeigne, France, has been regenerated by planting, including oaks (*Quercus robur*).

Neighbouring countries were much influenced and Swiss foresters, for example, adapted the practices to their conditions, leading to a 'boom in reforestation' (Ortloff, 1999). By the end of the 19th century parts of Germany are described as almost devoid of broadleaved forest, having been largely replanted with spruce (*P. abies*) and pine (*P. sylvestris*) (Jones, 1965; Kenk and Guehne, 2001) a conversion still being advocated in the early years of the 20th century (Leyendecker-Hilders, 1910).

Silviculture in central Europe was – and frequently still is – characterized by high levels of stocking, light thinnings, long rotations and a cautious attitude towards use of exotics. And, just as today, there were some misgivings about this kind of forestry (Jones, 1965) as well as enthusiastic supporters such as Leyendecker-Hilders (1910) in Germany. Simpson (1900) was so keen that he dubbed this silviculture as the 'new forestry' and urged British foresters to adopt such practices wholeheartedly. Interestingly, in the USA at this time even George Perkins Marsh, the 'prophet of conservation', averred that biblical stewardship both cares for and improves nature – that the artificial forest outshines the natural, the introduced crop outranks the native (Lowenthal, 2000).

Concurrent with the developments of new ways of growing trees was a rigorous evaluation of introduced species as collectors sent seeds and plant material to European countries from all over the world. With the minor exception of southern beech (*Nothofagus* spp.), all the principal exotic species so widely planted in the UK today (*P. abies* and *P. sitchensis*, *Pinus nigra* and *P. contorta*, *Pseudotsuga menziesii* and *Larix kaempferi*) were introduced more than 150 years ago. In warm temperate and Mediterranean regions this trial and testing of tree species led to widespread use of eucalypts, most notably *Eucalyptus globulus* and, even more importantly, to the phenomenal success of *Pinus radiata*, particularly in northern Spain and the four southern hemisphere countries of Australia, Chile, New Zealand and South Africa.

2.3.2 Tropical and subtropical countries

Development of tropical plantations can be traced back to the 16th and 17th centuries, with the expansion of European influence by the colonial powers. The colonizers encouraged an exploitative timber export trade, often seriously damaging natural forests. But until the late 1800s there was generally timber available and little need to plant trees in the tropics for industrial wood production. The scientific study of plants and animals led to systematic collections, botanical gardens flourished and the domestication of several tropical tree species began (Turnbull, 2002). Plantation activities before 1900 included the introduction and testing of exotic species, especially teak and eucalypts, and the introduction of taungya and irrigated plantations. The establishment of government agencies, use of trained foresters and definition of forest policies and legislation provided an institutional framework on which the extensive forest plantations of the 20th century could be based.

There is a long history of planting teak (*Tectona grandis*) in the tropics. In Asia it was extensively planted for timber in Java under the control of the Sultans in the 15th century. With the arrival of the Portuguese in the 15th century and the Dutch in the 17th century, the demand for the durable teak timber for general construction and shipbuilding intensified. By 1748 the Dutch East India Company controlled all teak forests and monopolized teak trading. Influenced by three German foresters, all the Javanese teak forests were brought under regular management in the late 1800s. The taungya regeneration method was introduced in 1873 and from 1895 almost all teak forests have been regenerated by this system. Taungya is a system in which farmers plant tree seeds or seedlings to make forest plantations and tend them in association with their food crops. The term 'taungya' originated in Burma (Myanmar) and its application in government forestry is usually attributed to Dietrich Brandis, a forest officer in charge of Burmese teak forests from 1856 to 1862. In Myanmar, teak planting began in 1856 and continued into the 20th century though, regrettably, most plantations were damaged during the Second World War. Teak was successfully introduced into Sri Lanka by the Dutchman van Rhede as early as 1680 (Perera, 1962) and, as teak was in short supply for shipbuilding in India during the early 1800s, the Collector of Malabar in 1840 suggested teak should be planted and the first plantations were established in the Nilambur hills (Puri, 1960). Between 1841 and 1855 some 600 ha were established and from 1840 it has been planted both within and outside its area of natural occurrence. The taungya method was generally used to establish plantations in Karnataka, Kerala, Uttar Pradesh, West Bengal and most parts of Assam and Tamil Nadu, although many forests were naturally regenerated.

As early as 1790 several eucalypts (*Eucalyptus* spp.) were planted in the Palace garden at Nandi Hills near Mysore, India and became the seed source for widely planted Mysore gum (*E. tereticornis*). In South America eucalypts were introduced in 1823 to Chile, and specimens of *E. robusta* and *E. tereticornis* in Rio de Janeiro, Brazil date back to 1825 (Jacobs, 1981). *Eucalyptus globulus* was one of the first eucalypts to be used for plantations. By 1900 it could be found in Europe (Italy, Portugal, Spain), Africa (Ethiopia, Kenya, South Africa), Asia (China and India) and South America (Bolivia, Chile, Colombia, Peru). At that time it was primarily planted for ornamental purposes or fuelwood. Eucalypts, which grew fast, provided fuel for the wood-burning locomotives in Brazil, East Africa, India and South Africa, but were inferior for sawn timber production. Other species, such as *E. robusta* and *E. tereticornis*, were planted in many countries and formed the base on which large industrial plantations developed in the 20th century, when eucalypts became the most planted broadleaved species in the world.

Plantations were started in South Africa in the latter part of the 19th century. The first wattle trees (*Acacia mearnsii*) were planted for tan bark in 1864 and the first pine (*Pinus* spp.) plantations in 1884. Pines were not commonly planted before 1900. *Pinus patula* was introduced to New Zealand in 1877 (Wormald, 1975) and probably other countries at about the same time, though it was not introduced into South Africa until 1907. *Pinus caribaea*, now widely planted in the tropics, was little known botanically, let alone planted, until 60 or 70 years ago. The lack of records of early introductions of pines may be because many quickly failed because of the lack of suitable mycorrhizas on the new sites.

Two important silvicultural practices saw their introduction during the latter part of the 19th century. Use of the taungya system to plant teak (*T. grandis*) has already been mentioned. The other was irrigated planting, which is generally associated with arid sites, where the annual rainfall rarely exceeds 200 mm, or semiarid areas with a very short rainy period. Irrigated plantations are usually close to major rivers. In the Indus basin of Pakistan, irrigation of *Dalbergia sissoo* and other species has been practised for more than 140 years. The first plantings were in the Punjab in 1866 to supply firewood for a new railway and fuelwood for Lahore and other cities. Notable irrigated plantations have subsequently been established in Iraq, Egypt and central Sudan. Species such as eucalypts, casuarinas and poplars are commonly grown in these irrigated systems.

Before 1900 there was no need to plant trees extensively as an industrial resource in the tropics, though several European countries were concerned about their own lack of natural forest. The main contribution of the pre-1900 period to tropical plantation silviculture today was the successful introduction and trial of several exotic species, notably teak and some pines and eucalypts, and the inception of taungya and irrigated plantations.

2.4 The Period 1900–1945

The first half of the 20th century saw the first extensive plantings of industrial tree crops, mostly in countries with little utilizable natural forest or, in the tropics and subtropics, where there had been an early influx of European settlers. The UK, at the time the world's largest timber importer and with only 5% forest cover in 1900, instigated an afforestation programme to double its area of forest, Denmark pursued tree planting as the principal tool of silviculture, and Australia and New Zealand began their massively successful *P. radiata* plantation programmes. Well-forested countries like the US embarked on major planting of southern pines (*P. elliottii* and *P. taeda*) to augment the industrial resource – indeed by 1942 Toumey and Korstian's substantial *Seeding and Planting in the Practice of Forestry* was already in its third edition, and the then USSR planted numerous shelterbelts to assist food production in the Ukraine and elsewhere.

In the tropics and subtropics, South Africa had, by 1945, created 180,000 ha of plantations of *P. patula*, *P. elliottii*, *P. taeda* and some *Eucalyptus* spp. in Mpumalanga (Eastern Transvaal) province alone, Queensland (Australia) had 9800 ha of exotic *P. elliottii* plantations and native *Araucaria cunninghamii* (Ryan and Shea, 1977) and in India, 80,000 ha of teak (*T. grandis*) had been planted (Griffith, 1942) and many trials of *Eucalyptus* spp. set up (Khan and Chaudhary, 1961).

Before 1945 the main plantations in tropical America were on a small scale for protective purposes around cities, and for fuelwood, railway sleepers and pit props. The most extensive early plantations were in the State of Sao Paulo, Brazil from 1905 to 1915 at the instigation of Edmundo Navarro de Andrade, head of forestry services for the Paulista Railway Company. In 1950 Brazil had close to half a million hectares of planted eucalypts, a larger area than any other country.

Between 1900 and 1945, although most of the world's tropical plantations for wood production were primarily of pines, eucalypts and teak, there were major plantings of trees for non-wood forest products. Paramount was the rubber tree (*Hevea brasiliensis*); first planted in Malaysia in 1898, it boomed in about 1910 when it was enthusiastically taken up by coffee and tea planters whose commodities were experiencing difficulties. Indonesia and Malaysia are currently the world's largest rubber producers. In the period 1920–1930, the private sector began extensive plantations of the Australian black wattle, (*A. mearnsii*) for tan bark to supply tannin to the leather industries. In 1921 there were about 115,000 ha in South Africa and 25,000 ha in Kenya. It was also planted in Zimbabwe, Tanzania, India and Brazil. In recent years the value of the tannin has declined relative to the value of the wood, which makes good charcoal and excellent paper pulp.

In the Indian subcontinent many pioneering developments in tropical silviculture took place and, as early as 1912, Broun described in detail afforestation and plantation practice. Indeed, India's tree planting programmes were well in advance of those of the then colonial power, the UK. However in most tropical countries forest plantations were still in their infancy, though many silvicultural developments took place that were later to feature strongly, most notable of which were: the introduction of the taungya system in many tropical countries; species and provenance trials; the development of the famous and revolutionary thinning schedules by Craib (1934, 1947) in South Africa, which has so impacted plantation silviculture elsewhere; enrichment planting to supplement inadequate natural regeneration; and 'compensatory plantations' to augment local wood supplies as areas of natural forest declined. Arguably, the best early example was the 4000 ha of *E. globulus* planted between 1900 and 1920 on the amphitheatre of hills around Addis Ababa, Ethiopia, which made up for the loss of indigenous forest in the previous decade. Wood harvested from successive coppice rotations of the current 15,000 ha of eucalypts has provided a sustained supply of fuel for the city to this day (Fig. 2.3).

Two further points should be noted for this period. First, the great bulk of tree planting was carried out by the state as a specific tool of forest policy. In many countries this was afforestation of bare land, e.g. in the UK, or conversion of what was deemed unproductive native forest, e.g. pines (*P. radiata*) replacing eucalypt woodland in Australia (Turner *et al.*, 1999).

The second point is the raising of serious doubts about the sustainability of plantation silviculture in the very country, Germany, where it had been so successful. Many second and third rotation spruce (*P. abies*) plantations, particularly in Saxony, were showing declining yields and other signs of ill-health, which were attributed to the effects of the clear-cutting and replanting silvicultural system (Weidemann, 1923). There was considerable alarm and attacks on the

Fig. 2.3. Woman carrying sticks and twigs from *Eucalyptus globulus* plantations above Addis Ababa, Ethiopia. Note coppice growth to the left of the donkeys.

concept of planted forests (Anon., 1923), and specifically that of conifer monoculture (Stickel, 1928), though many felt it was more to do with planting spruce (*P. abies*) 'off site' in the first place: the issues were never fully resolved (Evans, 1976). Similar concerns re-surfaced in Australia in the 1960s and are examined further in Chapter 6.

2.5 The Period 1945–1980

Three post-war trends can be identified for this period, which bear on the development, including expansion, of planted forests.

2.5.1 Internationalism and development

After the Second World War many new international agencies arose, including the UN FAO, which have had an important influence on forest policy and development. But, in addition to their general stimulus through provision of funds, development aid and technical expertise, substantial direct investment in industrial plantations also began. For example, both the major plantation projects in Swaziland date back to this period and were funded mainly by private companies with some institutional backing (Evans and Wright, 1988).

Declining colonial influence and the newly emerging independent states after the Second World War altered economic and developmental pressures in many tropical countries. Development became a politically important process: the key to meeting the people's aspirations in the new countries. Development of forest plantations can claim no great part in aid programmes of this period, but in several countries tree planting increased substantially as part of the overall development process. And, as King (1975) pointed out, almost every tropical country has at one time or another considered establishing industrial plantations for pulpwood.

Also in this period, several institutions were founded with regional responsibility or influence to assist forestry development, in particular plantation establishment, e.g. Centre Technique Forestier Tropical in many French-speaking countries, the Institute of Tropical Forestry in Puerto Rico and Centro Agronomico Tropical de Investigacion y Enseñanza in Costa Rica. Many university and college courses in forestry in developing countries were introduced at this time.

This increasing momentum was replicated in temperate countries by state or state-subsidized planting programmes with, for example, tens of thousands of hectares of new planting in Chile, New Zealand and the UK to name but three countries.

2.5.2 Awareness of silvicultural potential

As they matured, the earlier trials and older plantations demonstrated whether or not plantation silviculture was worthwhile. The general observation of superior growth compared with native species appeared true of many species on many sites, though the best matching of species or provenances with particular sites was not always achieved.

Evidence of the increasing awareness of the silvicultural potential of plantations – and also the new internationalism – is seen in the resolutions and subject matter of international conferences and meetings. At the Fourth World Forestry Congress (1954) in India, it was recommended that an international commission be set up on the use of exotic species for planting in the tropics. In 1957 the Seventh British Commonwealth Forestry Conference resolved that a book be published about experience with exotic species in the Commonwealth (Streets, 1962). In the 1950s both the International Poplar Commission and the Teak Sub-Commission were set up by the FAO with the great bulk of their work focused on plantation crops. The increasing importance of eucalypts for plantations was indicated by the publication by FAO in 1954 of the first edition of *Eucalypts for Planting* and the establishment of a 'Eucalypt Clearing House' service in 1962 by the Forestry and Timber Bureau in Australia to provide information and well-documented seeds for research and commercial plantations. And, according to Lamb (1973), the FAO Seminar on Tropical Pines in Mexico in 1960 more than anything else awakened tropical countries to the value of these species.

These initiatives and the gathering momentum of interest prompted what became the pivotal FAO *World Symposium on Man-Made Forests and their Industrial Importance* in Canberra in early 1967. The symposium, with participants from 41 countries and producing over 2000 pages of papers in three weighty volumes, testified to the increasing emphasis on plantations and their expanding role across the world. Many of the trends initiated in the 1950s and 1960s accelerated in the 1970s, new projects multiplied, afforestation became an important part of national forest policies, incentives were introduced to encourage private sector tree planting, and the importance of trees and forest in the environment became more widely recognized. In this period, Brazil's afforestation rate peaked at an astonishing 0.5 million ha per year, Australasia's, Europe's and Africa's largest man-made forests were all created, namely Kaingaroa in New Zealand, Kielder in England and Usutu in Swaziland, incidentally all with exotic conifers, and both China and India pursued increasingly ambitious planting programmes in successive development plans. In these latter two countries at least equal emphasis was given to the environmental roles tree planting could play. China sought to tackle desertification by planting *Populus* spp. in the temperate north and for dune stabilization using *Casuarina* spp., as well as continuing the massive Chinese fir (*Cunninghamia lanceolata*) afforestation in the central south of the country. In India one of the key programmes specifically targeted waste land afforestation.

Sometimes enthusiasm went too far. Throughout the 1950s and 1960s in the UK, some considered the great tracts of successful afforestation in the uplands too dense and unforgiving, while in the lowlands many mainly native broadleaved woodlands were converted to conifers on the grounds of economics. It was repeating German experience of a century before, but the practice was expensive, not entirely successful and later much criticized on landscape and conservation grounds. In the tropics numerous attempts to plant mahoganies (mainly *Swietenia macrophylla*) and other valuable hardwoods of the *Meliacaea* family failed because of poor silviculture and an inability to cope with *Hypsipyla* (*H. ferrealis* and *H. grandella*) shoot borer damage (Evans and Turnbull, 2004).

In many well-wooded countries planting came to be seen as a sensible handmaiden to achieve satisfactory regeneration. For example, in Sweden since the 1940s regeneration has relied on reforesting largely by planting (Savill *et al.*, 1997) and in the Pacific North-West of the USA clearcutting and replanting became the commonest way of regenerating Douglas fir (*Ps. menziesii*) (Smith, 1986).

Forest plantation development across the world became more important as the mid 20th century progressed and management practices intensified. It was realized that the genetic principles being applied to agricultural crops could be used to improve the productivity of trees in plantations. Forest geneticists in Denmark, Sweden and the USA led the way by elucidating patterns of variation in commercially important trees, and achieving basic understanding of pollination biology and vegetative propagation. Foresters in the tropics also recognized the need to improve seed quality of plantation species and some designated special seed collection areas. These advances paved the way for forest industries in the 1950s to launch with some confidence the large programmes of applied genetics to improve wood production in plantations (Turnbull, 2002).

Silviculturally the one uncertainty was a re-awakening of the question of the sustainability of yields over successive rotations. Alarming reports from Australia in the 1960s of declining productivity of second rotation *P. radiata* (Keeves, 1966; Bednall, 1968) led to much research, but incidences elsewhere were sporadic (Evans, 1999). Indeed, Holmsgaard *et al.* (1961), following up in Denmark with similar crops to the German reports of 30 years before, could find no evidence of yield decline. Nor was this the case for second rotation *P. patula* in Swaziland (Evans, 1978).

2.5.3 People, funding and environmental issues

Much of the expansion of afforestation was for industrial purposes – pulpwood, sawtimber, and to a small extent plywood veneers. But planting also increased to meet direct human needs (firewood, shelter, building poles, fodder for grazing animals) and for environmental protection. Precise figures are rarely available, but nearly every tropical country greatly expanded the supply of tree seedlings for extension purposes – the distribution of trees to farmers and villagers usually free of charge. This is clear from plantation data. By 1980, of 11.5 million ha of plantations established in tropical countries, well over a third (4.3 million ha) were for non-industrial purposes (FAO, 1988).

These wider objectives for tree planting were seen in the themes of the World Forestry Congresses 'The forest and socioeconomic development' (1972), and in 'Forests for people' (1978). Literature similarly reflected the changing emphasis, e.g. *Forestry for Rural Communities* (FAO, 1978), *Trees, Food and People – Land Management in the Tropics* (Bene *et al.*, 1977), and *Forest Energy and Economic Development* (Earl, 1975).

In the 1970s many organizations came to play a part in encouraging tree planting in the tropics, particularly with non-industrial projects. These were often the development of village woodlots, agroforestry or tree planting for environmental protection purposes. Policies of development banks were also revised specifically to include plantations for environmental as well as industrial purposes. In 1978 the World Bank stated that 60% of future lending would be for rural forestry projects, primarily to encourage village development, and only 40% to help finance large-scale industrial plantations.

Private investment in plantations increased, stimulated by favourable taxation, subsidies and related fiscal arrangements. In the UK investment companies were established specifically to harness these benefits and private sector planting increased five-fold (Grundy, 2005) and in the USA non-industrial private forest owners (NIPF) were planting a remarkable 400,000 ha/year by the late 1950s under the Soil Bank Program (Alig *et al.*, 1999; Fox *et al.*, 2007).

Forest plantation development expanded rapidly in all aspects. However, while this period saw expansion, and in several countries a switch to it from attempts at natural forest management, there were failures. Valuable plantations were neglected or abandoned, examples from Africa include Cameroon, Gabon, Liberia and Zaire, owing to budget cutbacks and an inability to meet the expenditure required to maintain the resource. And, equally regrettably, many early attempts at social forestry in the tropics were unsuccessful often for sociological rather than technical reasons, such as failure to engage adequately with key stakeholders.

In developed countries increasing public demand for recreation, amenity and enjoyment of the outdoors began to modify plantation practices. The industrial plantation had to serve more than the one commercial objective: issues of landscape and aesthetics, tree sizes and rotation lengths, and even species choice – the professional forester's core competence – began to be questioned. And it was found that planted forests could do almost as well as natural forests in meeting these wider needs if carefully managed. The public at large became stakeholders of forestry and especially forest plantation practices.

2.6 From 1980 to the Present

For the immediate past, this summary of the history of planted forests is briefer because it is better known and because information is readily available from the proliferation of literature, websites and communication generally. Only highlights will be given for six areas: industrial plantation resource; plantings for domestic and non-wood products; planted forests and rehabilitation, conservation and environmental roles; carbon storage and related climate change issues; the enjoyment of planted forests for recreation, amenity and their wider impact in the landscape; and an increasingly diverse and vocal stakeholder community. These distinctions are artificial and it bears restating, as was noted in Chapter 1, that 'All planted forests can offer several, even many, benefits'. Achieving these benefits and complying with good standards have become, in recent years, increasingly the preserve of certification bodies. Also, many issues raised here are re-visited in Chapter 6.

2.6.1 Industrial plantation resource

Data show the continued expansion of forest planting worldwide. Increasingly the view took hold that planted forests were destined to become the world's principal 'fibre basket'. In many countries earlier programmes continued apace, e.g. Chile, India and the USA (Prestemon and Abt, 2002; Alig and Butler, 2004), and in some, such as China, expanded massively (Zhang and Song, 2006). But in other countries, afforestation – planting forest to increase forest cover – began to tail off and be replaced by planting to restock after felling, e.g. New Zealand, South Africa and the UK. In contrast, some forest-rich countries with little history of planting now saw the practice gain in importance. They are well illustrated by Brazil's massive investment in *Eucalyptus* and *Pinus* plantations, by Indonesia establishing extensive plantations of exotic species such as *Acacia mangium*, *Eucalyptus grandis*, *P. caribaea* and *Gmelina arborea*, mostly to replace cleared natural forest, and by Canada where, from the 1980s, planting increased massively (Reed, 1983; Lavender, 1990; Binkley, 1999; Binkley *et al.*, 2005), but almost exclusively of native species, such as fir (*Abies* spp. and *Pseudotsuga menziesii*), hemlock (*Tsuga canadensis*), pine (*Pinus* spp.) and spruce (*Picea* spp.), as a way of ensuring regeneration and ultimately securing future production.

Increased recourse to planting is seen elsewhere. Planting, mainly with native Douglas fir (*Pseudotsuga menziesii*), was the principal means of recovering forest around Mt St Helens, Oregon, USA following the volcanic eruption of May 1980. While much of the new planted forest will be reserved as a national park, much will provide future lumber. In central France the storms that wrecked tens of thousands of hectares of oak, beech and fir forest in the mid-1990s, have mainly been replanted. And around the Indian Ocean, following the terrible tsunami of December 2004, planting programmes are beginning to reinstate many of the uniquely valuable mangrove forests, the loss of which contributed to the scale and severity of the devastation.

Both intensification of management of industrial plantations and the converse have taken place. Most forest plantations for industrial products consist of genetically improved material, ranging from better provenance, site matching to clonal crops and increasing use of biotechnology (Evans and Turnbull, 2004; Johnson and Kirby, 2004; Nehra *et al.*, 2005), and incorporate establishment and protection procedures to maximize yield and crop health. This intensification is not confined to the growing stage. Increasingly wood-using industries began taking smaller size material, which plantations are well able to supply, for reconstituted wood products. At the same time, many extensive plantations began to be managed in ways that did not focus exclusively on maximum fibre yield, but to accommodate other interests and priorities.

2.6.2 Non-industrial tree planting

From the early 1980s 'domestic' wood products such firewood and charcoal, building and fencing materials, and non-wood products such as foods and medicines increasingly became objectives for tree planting. While these were not new, the oil crisis of the 1970s focused concern on all fuels and the realization that most developing countries remained massively dependent on wood for their energy. Many woodlots and village-scale plantings first date to this period and many, as mentioned earlier, were often unsuccessful. But what has been successful, as these kinds of planting have grown in significance, is engagement with local people's needs and aspirations as exemplified by India's participatory programmes (Saxena, 2001). A large literature now exists for rural development forestry.

A parallel development, though far less significant in terms of meeting people's basic needs, was the rise of 'energy' plantations. Although this term was being used in the 1970s (e.g. Evans, 1976), it is more recent times, as worries about fossil fuels grew, that serious research and significant investment has been made into growing wood as an energy feedstock. This has been widespread, even in forest-rich countries like Sweden. Concerns of land availability have been raised (Nonhebel, 2005).

Under this heading can be included agroforestry – combining the growing of trees and food in various ways (Sinclair, 2004) – and all tree planting outside forests, both of which gained enormously in significance.

2.6.3 Forest rehabilitation, conservation and environmental imperatives

Any forest, including planted forest, possesses a more equable microclimate than that of most farm land and urban development. This long-recognized forest influence began to be used to assist rehabilitation and restoration of former forest land that had been cleared in the past. The main aim was not to establish a plantation *per se* for its products, but to harness tree planting to re-introduce woodland conditions conducive to the recovery of natural forest. There are now many examples in the tropics where natural forest once converted to ranch or range land is returning to its former condition, facilitated by a stage of tree planting (see Fig. 2.4). Evans and Turnbull (2004) devote a whole chapter to this topic. And there are many examples of planting indigenous species to initiate the recovery process, even on land that has been farmed for centuries.

The converse of this has been the deliberate removal of planted trees, especially of exotic species, from former native woodland sites to help reinstate the latter. This is funded in the UK under the PAWS scheme (Plantations on Ancient Woodland Sites) and seeks to reverse much of the ecological 'damage' caused by the excesses of the 1950s and 1960s noted earlier (Harmer and Kiewitt, 2006). Related to this were increasing attempts to restructure uniform planted forests at the landscape level (Hibberd, 1985; Evans and Hibberd, 1990; Gill and McIntosh, 2001) and at the stand level through processes of 'transformation' (see Fig. 2.5). Continuous cover forestry (CCF) matches traditional shelterwood and selection silvicultural systems, and brings 'naturalness' to maturing plantations through extended regeneration (Cameron *et al.*, 2001).

And, added to these direct impacts of trees, including planted trees, is a wider recognition of conservation imperatives, the need for operational guidelines to minimize impact on soil and water, sites of archaeological interest and the

Fig. 2.4. Planted *Cordia alliodora* in Ecuador on former ranch land initiating return to native forest conditions.

landscape generally. Again, to cite the UK, these issues began to take precedence over economic objectives and tree planting today only attracts grant aid if it delivers these non-wood and often non-market benefits.

2.6.4 Carbon storage and climate change issues

Trees accumulate carbon during their life and can often enrich soil surface carbon stocks. Thus appropriate afforestation was seen as one of the weapons to slow the rise in atmospheric carbon dioxide and in turn help mitigate global warming. Inclusion of tree planting in the Kyoto Protocol acknowledged this role (Bloomfield, 2000) – while recognizing in global terms that it was far more important to prevention deforestation – but deliberate afforestation to offset carbon emission or as part of carbon trades was slow to develop. This partly arose from

Fig. 2.5. One of Europe's largest planted forests, Kielder Forest, England, being restructured at stand and landscape level to increase diversity. Uniform stands of mainly *Picea sitchensis* are deliberately broken up to widen age range, diminish block size, add species and in places, restore wet mires.

the recognition that some kinds of tree planting were not beneficial in this way: (i) where natural forest formations were cleared simply to allow plantations to be established, which almost always has a negative net carbon sequestration impact; and (ii) where plantations were established on peats and other carbon-rich soils and operations such as ploughing accelerated the breakdown (oxidation) of organic matter and liberated more CO_2 than the planted trees would ever store. Of course, using afforestation in these ways could well bring conflicts of other kinds, particularly social (Smith and Scherr, 2003).

The issue was further aggravated by claims of many NGOs, and with some justification, that massive tree planting to offset carbon use diverted attention away from the pressing need to reduce wasteful per capita consumption of carbon, such as fossil fuel energy. The Expert Meeting in Chile on the Role of Planted Forests (Anon., 1999) made this point explicit in Recommendation (iii).

2.6.5 Recreation, amenity and the wider landscape

Planted forests have a dramatic impact on the landscape owing to their orderliness and uniformity. Their internal landscape of sharp edges to compartments, straight rides and breaks, and monoculture of species is similarly unattractive, if economically efficient. Yet these impacts are what many of the public see as forest. With sympathetic management the starkness can be mitigated and the forest, though originating by planting, becomes close to natural in feel and perception, if not in reality. These issues, like those of conservation, greatly influence practices and even policies to the point of denial that forest has a planting origin.

2.6.6 Wider stakeholder community and conflict

The above remarks indicate a far wider public interest in plantations, especially the impact of large-scale tree planting. Concern ranges from the accusation of plantations being 'green cancer' in the landscape to insistence of public access wherever public funds (taxes) are spent. Compared with previous periods in the history of planted forests, there are many more stakeholders expressing a view and claiming an interest. Conflicts of interest, particularly over threats to biodiversity, social and community issues, water resources and landscape impact, have been many. Books such as *Plantation Politics: Forest Plantations in Development* (Sargent and Bass, 1992) and *Towards a Sustainable Paper Cycle* (IIED, 1996) are illustrative of heightened debate. The following are examples by country.

- Australia – community concerns about ownership and related socio-economic issues (Barlow, 2003).
- Canada – plantations and biodiversity (Betts *et al.*, 2005).
- Chile – biodiversity and indigenous people (Armesto *et al.*, 2001) and water resources (Pizzaro *et al.*, 2006).
- China – water resources (Sun *et al.*, 2006).
- India – water resources, eucalypts and local people (Evans and Turnbull, 2004).
- Indonesia – local community conflict with companies (Nawir and Santoso, 2005).
- USA – environmental impacts (Hayes *et al.*, 2005), public engagement (Howe *et al.*, 2005) and biodiversity issues (Carnus *et al.*, 2006).
- UK – water resources (Calder *et al.*, 2003).

Addressing these issues of conflict has led to several responses, but overall it is through the tool of criteria and indicators, developed in the 1990s following Rio 1992, and the policing of their compliance and of related issues through certification processes that have been some of the most significant developments in recent years. Other examples of responses of conflict resolution include:

- Well-publicised conferences, e.g. Planted Forests Symposium in Oregon in 1995 (Boyle *et al.*, 1999).
- The need to demonstrate genuine sustainability (Brundtland, 1987) and questioning the moral obligations surrounding intensive management of plantations (Moore, 2005).
- Reviews undertaken by large corporations, often in association with environmental NGOs, e.g. Shell/WWF Tree plantation review in 1993 (Adlard, 1993).
- Rapid uptake of certification in forest plantation management (Leslie, 2004; Gouldin, 2006) to demonstrate, amongst other things, compliance with the best environmental and conservation practices – the bulk of both

South Africa's and the UK's forests are certified and they are dominated by forest plantation.

- Greater emphasis on native species.
- Confining use of public funds for tree planting to non-market benefits.

2.7 Some Conclusions

This outline of the history of planted forests and issues that have arisen draws out helpful pointers for the future.

- Tree planting has a long history: it is a very old practice.
- Reasons for planting forests were frequently prompted by local or regional shortages of wood, fuel and lumber.
- While afforestation – creating new forest by planting – has always featured, much early use of tree planting was to augment inadequate or missing natural regeneration and enhance productivity. Tree planting ensured satisfactory restocking of existing forest.
- Conversion of forest from one species to another through enrichment or by wholesale replanting also has a long history, but not always one of success in the long term.
- Use of exotics has resulted in some of world's most productive forest plantations.
- Concerns over sustainability, and more generally the use of monocultures, go back at least 100 years, but have not seriously challenged plantation silviculture as a practice.
- While much tree planting has had production or commercial objectives, planting for protection of soil or assisting conservation was practised from an early time. Planting forests for multipurpose use has been commonplace.
- There has been a long history of tree planting outside the forest, including agroforestry, to provide goods and services and to combine food and tree growing in socially acceptable ways as part of forest establishment, e.g. taungya.
- Some plantation practices have intensified to increase productivity per hectare, while at the same time there has been recognition that management over whole forests must be environmentally sensitive, in the interests of landscape, amenity, conservation and related imperatives.
- Concomitant with the above has been the rise of stakeholder involvement, both to influence planted forest development and to engage in tree and forest planting itself.

3 The Question of Definitions

J. Evans, J.B. Carle and A. Del Lungo

3.1 Introduction

Definitions and interpretations are always contentious, for three reasons. First, what is actually understood by a word – what does it mean and has its meaning changed over time or is its meaning slightly different from one culture to another? Second, how is something classified to fit a certain definition – what are the features that make it so? And, third, does the classification apply to the whole of what is being described or, if only a part, what proportion is sufficient to meet a particular class? It's the problem of how to cope with a continuum of variation. Abstract though this is, it has beset the use of words like 'plantation', 'forest plantation' or its reverse 'plantation forest', and 'planted forest'.

Two hundred years ago a 'plantation' was what slaves laboured in to grow sugar or where rows of trees were planted on bare land. Today the slaves have gone and the rows of trees have disappeared as thinnings, fellings and regeneration have led to natural-looking forest with little sign of its origin by planting. Does the passage of time change the classification?

Planted forest is the label used to overcome this issue, embracing both the new and readily understood 'plantation' and those forests which were originally planted but now no longer appear so.

3.2 Coping with the Continuum

3.2.1 Many roles and many types of planted forest

The history of tree planting reveals not only how old the practice is and its recent and substantial expansion, but also the diversity of where planting is done. While the act of planting trees defines the subject, there is a continuum of types and the interface between some planted forests and natural regenerating forests is indistinct. Trees and forests cannot simply be divided into those arising by natural means and those that have been planted or directly sown. There is the continuum from untouched natural forest to tree cropping – the image most associated with a plantation. And, as touched upon, there is a continuum from those planted long ago but now seeming wholly natural to recent afforestation of bare land. A yet further continuum is from a forest of exclusively native species to one exclusively exotic – but when did plant distribution comply with geopolitical boundaries, or what about an exotic plantation with an understorey rich in native shrubs? Other continuums can be considered, and each has a midrange where things are indeterminate. But lack of agreement on interpretations of the definition of 'plantation' or 'planted forest' – where the cut-off points are on the continuum – causes problems in comparing forest resource statistics from different sources. Thus definition is important, not simply for gathering accurate statistics *per se*, or to improve communications, or even to allow better production forecasts, but because of the perceived attributes of and common preference for the natural in terms of landscape, amenity and biodiversity.

We need to go further. Trees, woodland and forests offer many social, ecological and environmental functions, but so can horticultural crops, oil palm and rubber plantations. Functionality as well as description is critical. Indeed, Adlard (1993), in an analysis of types of plantation, identified no less than seven ways of categorizing them by: the object of afforestation; what the former land use was; who the key decision makers were; how the plantation fitted with agriculture; what the plantation contributed to conservation and environmental improvement; the types of produce; and the social benefits. The matrices were more a way to classify types than to define terms.

These complex and inter-related issues have led to many calls for clarity of definition.

- The question of definitions was raised at both UNFF intersessional meetings on the Role of Planted Forests in Sustainable Forest Management in Chile (Anon., 1999) and in Wellington, New Zealand (Anon., 2003).
- The web site of Lund (2000) provides a plethora of definitions from many countries and illustrates the complexity of the problem.
- It was the subject of a typology review in 2002 (Poulsen, 2002; and see Table 3.1).
- FAO has taken a lead on harmonizing forest-related definitions (FAO, 2003).

Each of FAO's Forest Resources Assessments have adopted variants of definition and attempting to make clear the distinctions in the continuum for purposes of reporting relevant statistics: they can be found in the appropriate quinquennial report, but see also (FAO, 2003).

3.2.2 Evolution of definitions

The pivotal 1960s conference on man-made forests (FAO, 1967) adopted origin as the basis of classification of broad forest types. And with minor variants this remained in place for the next 30 years.

Table 3.1. CIFOR typology of forest plantations and definitions.

Typology	Description
Industrial plantation	Intensively managed forest stands established to provide material for sale locally or outside the immediate region, by planting and/or seeding in the process of afforestation or reforestation. Individual stands or compartments are usually with even age class and regular spacing and: • of introduced species (all planted stands) and/or • of one or two indigenous species • usually either large scale or contributing to one of a few large-scale industrial enterprises in the landscape.
Home and farm plantations	Managed forest, established for subsistence or local sale by planting and/or seeding in the process of afforestation or reforestation, with even age class and regular spacing. Usually small scale and selling, if at all, in a dispersed market.
Environmental plantation	Managed forest stand, established primarily to provide environmental stabilization or amenity value, by planting and/or seeding in the process of afforestation or reforestation, usually with even age class and regular spacing.
Managed secondary forest with planting	Managed forest, where forest composition and productivity is maintained through additional planting and/or seeding.
Managed secondary forest without planting	Managed forest, where forest composition and productivity is maintained through natural regeneration processes, which can include the use of seed trees.
Restored natural/secondary forest	Restored forest, through either planting and/or seeding, or through natural regeneration processes, where restoration aims to create a species mix and ecology approaching that of the original natural forest.

However, planted forests may be classified in several ways, e.g. by their species composition, their scale, their complexity, the functions they perform or the purpose for which they are planted (Evans, 1999). Before FRA 2005, for definitions relevant to resource assessment see FAO (2001) and Carle and Holmgren (2003). (For a discussion of definitions relevant to climate change and carbon sequestration see Noble *et al.*, 2000). The key issue that became the focus of definitions was that of management intensity. The point being that although planted forests are frequently managed intensively for wood production, they can also be managed less intensively for conservation, protection or other socio-economic purposes. Thus FAO (2001) defined plantation as: 'Forest stands established by planting or/and seeding in the process of afforestation or reforestation. They are either:

- introduced species (all planted stands), or
- intensively managed stands of indigenous species, which meet all the following criteria: one or two species at plantation, even age class, regular spacing.'

While this definition covered most situations, there remained the question of the meaning of 'intensively managed'. It was intended to exclude stands established as plantations but now considered semi-natural because they have not been managed intensively for a significant period. The FAO definition was not intended to replace existing national classifications because national inventories, terms and definitions have specific purposes of relevance for each country (FAO, 2001). There are many types of plantations, ranging from short-rotation industrial tree crops through to 'close-to-nature' plantations which vary in intensity of management and other management practices according to whether the objectives are to maximize wood production, environmental values, or some combination of production and conservation objectives. Close-to-nature plantations are complex production systems using more than one species that may be unevenly aged, and several management practices, such as a mixture of coppice and standards, to provide a range of products and environmental services (Kanowski, 1997).

Thus the boundary between planted and natural forests can become very imprecise indeed. While consideration of origin is crucial, between the extremes of afforestation and unaided natural regeneration of indigenous forest, there is a range – a continuum – of forest conditions where intervention occurs to a greater or lesser extent in regeneration. In older typologies – systems of definitions – four broad forest types were, firstly, identified by origin.

1. Afforestation is the act or process of creating forest land where it 'historically' did not exist (Lund, 2000). Others have specified the time factor as 'where there has been no forest for at least 50 years' (FAO, 1967) or 'which previously did not carry forest within living memory' (FAO, 2001). Afforestation of grasslands falls into this category (Fig. 3.1) as does planting to stabilize sand-dunes.
2. Reforestation is the act or process of changing previously deforested lands back to forest land (Lund, 2000). A distinction can be made on the basis of whether the previous crop is replaced by the same or a different crop. An example of the latter is where rain forest is logged, cleared and then part replanted with a single tree species such as *Acacia mangium* or *Paraserianthes falcataria* (Fig. 3.2). The former is less common in the tropics, though *Araucaria cunninghamii* plantations in Queensland, Australia, some *Triplochiton* plantations in West Africa are examples. Common in temperate countries are oak (*Quercus robur* and *Q. petraea*) plantations in France, Germany and the UK and Douglas fir (*Pseudotsuga menziesii*) planted in the USA's Pacific North-West region and western Canada, where the previous forest is replaced by essentially the same species (Figs 2.2 and 3.3). Replanting was defined as the re-establishment of planted trees, either because afforestation or reforestation failed, or the tree crop was felled and regenerated (FAO, 2001).
3. Forests established by natural regeneration with deliberate silvicultural intervention and manipulation.
4. Forests that have regenerated naturally without human assistance, e.g. most natural forests in the tropics.

Up until FRA 2005, plantations are the forest types in classes 1 and 2 above, that is, artificial regeneration is the basic criterion.

Fig. 3.1. Afforestation of grassland with *Pinus patula*. Usutu Forest, Swaziland.

Fig. 3.2. Reforestation with a different species: *Paraserianthes falcataria* planted on recently cut-over rain forest in the Philippines.

Second, it was common to further differentiate between 'industrial plantations', which are established totally or partly to produce wood and fibre for industry (mainly saw-logs, veneer logs, pulpwood and mining timbers) and 'non-industrial' plantations established for one or more of the following objectives: fuelwood, wood for

Fig. 3.3. Reforestation by planting with essentially the same species: Douglas fir (*Pseudotsuga menziesii*). Pack Forest, Washington State, USA (planted semi-natural – Table 3.2).

charcoal, wood for domestic consumption, non-wood forest products and soil and water protection.

Third, in addition to their origin and the distinction between industrial and non-industrial, several other characteristics were presented as 'associated with' plantations. These include:

- Well-managed plantations usually have higher yields of useful wood than natural forests. Many commercial plantations, whether in temperate regions or the tropics, have an annual growth rate in the range of 4–40 m^3/ha compared with 1–10 m^3/ha for natural forests on equivalent sites, i.e. productivities typically two to four times greater.
- Plantations produce wood quickly and of a more uniform size and quality than that from natural forests. This facilitates harvesting, transport and conversion.
- Plantations can be located wherever infrastructure and suitable land are available, and near to population centres or wood processing units, thereby making them more easily accessible and reducing transport costs.

Finally, it was noted that traditionally trees had also been planted for landscape enhancement, prevention of soil erosion, providing shelter against heat and wind, and in agroforestry systems. Moreover, billions of trees planted in cities, on farms and along roads were not included in these definitions of 'plantations' and 'forests'. These 'trees outside the forest' were not included in plantation statistics but nevertheless make a substantial contribution to the environment and provide significant social and economic benefits. In Kenya trees on farms produced almost 10 million m^3 of wood in 2000 and in some densely populated areas contributed 18–51% of total household incomes (FAO, 2001).

The above definitions led to a complex of inter-related terms that resulted in fuzzy thinking, confusion over meanings and lack of transparency.

3.2.3 Proportions

What has not been addressed is the further issue of proportion. Many forests have a mix of silvicultural histories and treatments. Parts regenerate naturally, parts are planted, parts are enriched by planting and so on. Usual guidance is to classify based on whether more than 50% by area or number of trees falls in the category concerned. The difficulty is that such information is often unavailable.

3.3 FRA 2005 and the Planted Forests Subgroup

Since 1980 the Forest Resources Assessment (FRA) has been collecting statistical information on forest according to the two main classes of forests: natural forests and plantations. But in 2005 the Global Forest Resources Assessment (FAO, 2006a) introduced, for the first time, two additional forest classes, namely 'modified natural forests'[1] and 'semi-natural forests'.[2]

The new forest classification, composed of primary forests, modified forests, semi-natural forests and forest plantations, is based on different degrees of human intervention and method of regeneration. However, while it is easy to classify primary forests and forest plantations (and the modified forests, despite human intervention, can be considered as naturally regenerated forests), the semi-natural forests are a wider class and more difficult to classify owing to a dichotomy. They include components of (a) naturally regenerated and (b) planted forests, the latter similar to forest plantations in their use, their silvicultural practices and the intensity of management.

Table 3.2 shows, within the FRA forest categories, the links with the planted forest subgroup, namely, the planted component of semi-natural forests plus the productive and protective forest plantation.

The logic of this grouping is that the planted component of semi-natural forest, with intensive silvicultural treatments, is not materially different from forest plantations (FAO, 2006c). The only distinction is they are of native species and continue the character of the previous forest, i.e. fall in category 2 of the older typologies (page 26) but of the subset of where the previous crop is replaced by essentially one of the same kind. As the review of history records (Chapter 2), this planted forest type is very common. One of the earliest reasons for regeneration of forest by planting was simply to ensure satisfactory restocking and perpetuation of the forest.

The next two chapters examine the impact and the implications of this better way of classifying what is and is not planted forest.

Table 3.2. Planted forests sub-group in the continuum of FRA 2005 categories.

			Planted forests subgroup		
				Plantation	
Primary	Modified natural	Semi-natural		Productive	Protective
Forest of native species, where there are no clearly visible indications of human activities and the ecological processes are not significantly disturbed	Forest of naturally regenerated native species where there are clearly visible indications of human activities	**Assisted natural regeneration** through silvicultural practices for intensive management • Weeding • Fertilizing • Thinning • Selective logging	**Planted component** Forest of native species, established through planting, seeding, coppice	Forest of introduced species and in some cases native species, established through planting or seeding, mainly for *production of wood or non-wood goods*	Forest of native or introduced species, established through planting or seeding, mainly for *provision of services*

(FAO, 2006c)

Notes

1 Modified forests are 'Forest/other wooded land of naturally regenerated native species where there are clearly visible indications of human activities'. This definition includes, but is not limited to: selectively logged-over areas; areas naturally regenerating following agricultural land use; areas recovering from human-induced fires; areas where it is not possible to distinguish whether the regeneration has been natural or assisted.

2 Semi-natural forests are 'Forest/other wooded land of native species, established through planting, seeding or assisted natural regeneration'. This definition includes areas under intensive management where native species are used and deliberate efforts are made to increase/optimize the proportion of desirable species, thus leading to changes in the structure and composition of the forest, with possible presence of naturally regenerated trees from other species than those planted/seeded. May include areas with naturally regenerated trees of introduced species, and areas under intensive management where deliberate efforts, such as thinning or fertilizing, are made to improve or optimize desirable functions of the forest. These efforts may lead to changes in the structure and composition of the forest.

4 The Global Thematic Study of Planted Forests

J.B. Carle, J.B. Ball and A. Del Lungo

4.1 Introduction

The new concept of 'planted forests', which was defined in the previous chapter, brings together two broad kinds of forest that were formerly considered separately: plantations and the planted component of semi-natural forests. Together they constitute the planted forests subgroup. As noted, the main reason is that the planted component of semi-natural forests has much in common with plantations, owing to their usually being managed more intensively in the same way that plantations are, namely:

- Similar types of planting stock, frequently of improved germplasm.
- Similar methods of establishment and tending.
- Thinning and pruning.
- Outputs that are uniform in size and technical specification, and frequently of wood or fibre for industrial use.

Because of this new grouping and because planted forests reflect a higher social, environmental and economic importance than their area would suggest, FAO embarked on a Global Planted Forest Thematic Study (FAO, 2006c). To measure and plan the quantity and quality of planted forest resources and the provision of goods and services that they supply, detailed data were sought not only about areas, but also about species, growth rates (mean annual increment – MAI), rotation lengths, harvest yields, end use, ownership, age class distribution and new planting. This in-depth examination of the whole planted forests subset included both productive and protective functions while still distinguishing between them. The data allow analyses of the regional and global status of planted forest development and present trends, will assist formulation of policies and outlook studies, and improve planning, monitoring and reporting of this increasingly important subset. The grand aim is to provide a realistic picture of the status and trends of wood, fibre, biomass, non-wood forest products and services from planted forests.

4.1.1 The thematic study

The approach of the study was to supplement the FRA 2005 information (FAO, 2006a) by seeking the above additional data from all countries reporting significant areas of plantations and semi-natural forests. They included those with the 30 largest forest plantation areas reported in FRA 2000 (FAO, 2001), along with countries in Europe and North America believed to have significant planted components of semi-natural forests: in total, this amounted to 61 countries. However, not all countries responded, or were able to provide data for the new category. In these cases a desk study of published information provided the estimates.

Taken together, country responses accounted for 72% of global forest plantation area and 83% of the planted component of semi-natural forests. Combined with the desk study it is believed that close to 95%, or almost all, of the world's planted forest area is included.

Data were gathered for three years, 1990, 2000 and 2005, and at two levels: (i) areas and types of planted forests, based on country responses and the desk study; and (ii) the in-depth data noted above based on country responses only. Figure 4.1 illustrates this.

The data and statistics in this chapter are not all equally reliable. This arises from the question of definitions and their interpretation, from the availability of data or otherwise, and from not being able to survey all countries. In consequence,

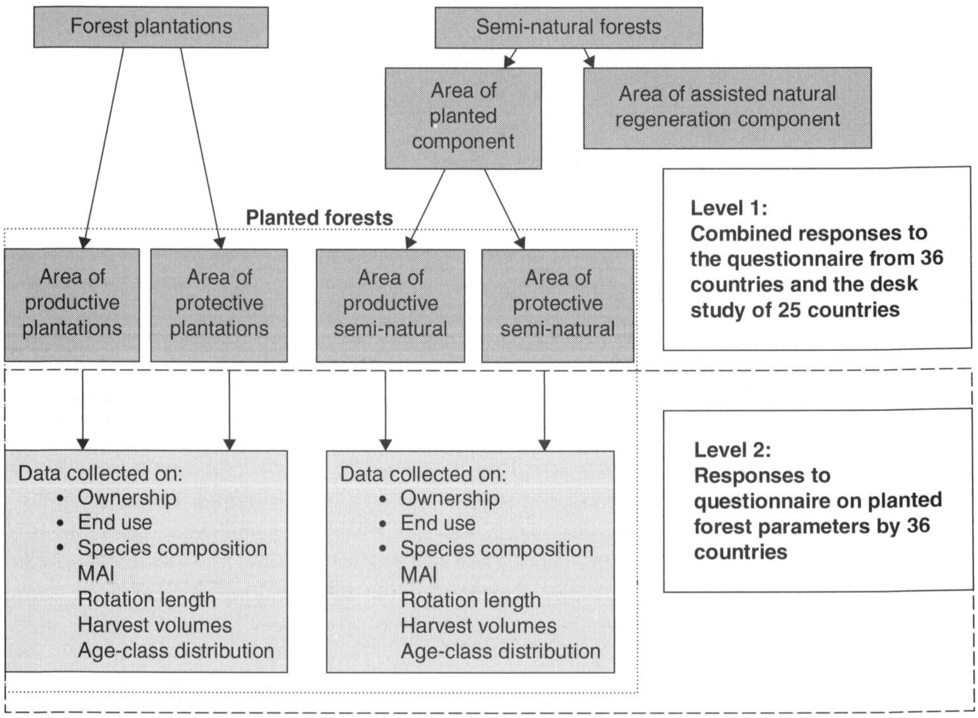

Fig. 4.1. Information requested and collected in the questionnaire and in the desk study.

data quality are believed: (i) robust, for gross areas; (ii) reasonable, for the distinction between 'productive' and 'protective' function; but (iii) only 'acceptable' for the in-depth analysis of productivity, ownership and related issues and thus useful only for indicating orders of magnitude and trends but not absolute figures.

4.2 The Extent of Planted Forests

This section brings together the results of the Global Planted Forest Thematic Study (FAO, 2006c) and the FRA 2005 (FAO, 2006a) data to estimate the total area of planted forests worldwide. The planted forest component of semi-natural forests is considered first, followed by forest plantation data from FRA 2005, but with some amendments.

4.2.1 Planted forest component of semi-natural forest

About half of all forests reported as semi-natural were allocated to the new category of 'planted'. And, as Fig. 4.2 indicates, this proportion seems to be rising with time and suggests preference is increasingly being given to planting as a way of ensuring restocking.

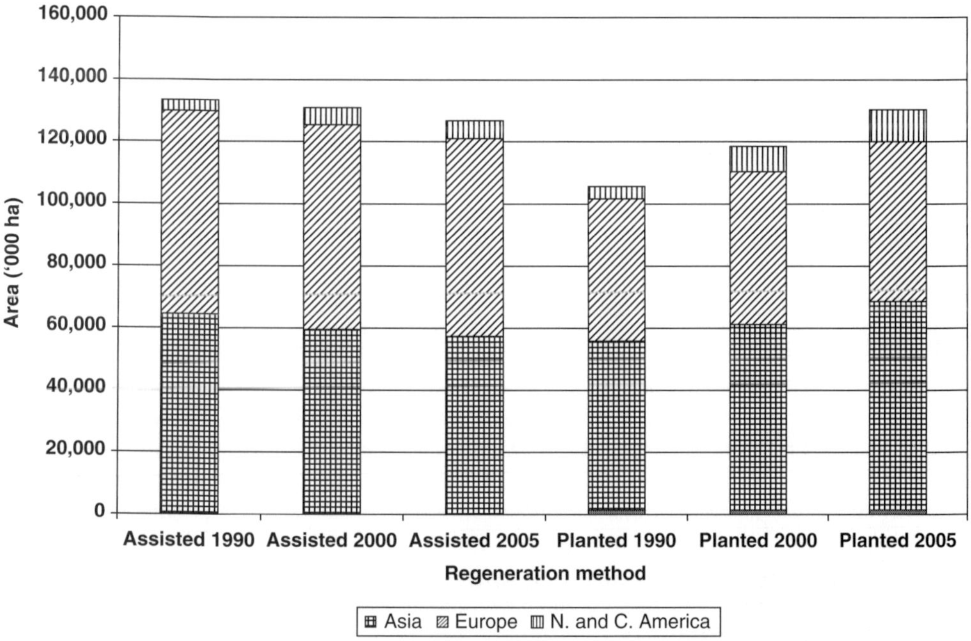

Fig. 4.2. Semi-natural forests by regeneration method by region, 1990–2005: 'assisted'[1] and 'planted' components. Little forest was reported in this semi-natural category from Africa, Oceania and South America.

Table 4.1. Areas of planted component of semi-natural forests by productive and protective functions: questionnaires and desk studies ('000 ha).

	Semi-natural 1990		Semi-natural 2000		Semi-natural 2005	
Regions	Productive	Protective	Productive	Protective	Productive	Protective
Africa	1,044	494	1,003	504	963	538
Asia	36,027	18,277	37,822	21,333	41,758	25,338
Europe	36,652	9,218	39,820	9,919	41,363	10,062
North and Central America	3,976	0	8,147	0	10,206	0
Oceania	0	0	0	0	0	0
South America	25	0	25	0	25	0
World	*77,724*	*27,990*	*86,817*	*31,756*	*94,315*	*35,938*
	105,714		118,573		130,252	

The surprising statistic is the total. The planted component of semi-natural forests, estimated at 130 million ha in 2005, is greater than the 110 million ha for all productive plantations in FRA 2005 (FAO, 2006a) and only a little less than the total of all plantations (Table 4.2 below). Less surprising is that Asia and Europe account for 90% of where planting in semi-natural forests is classified as planted forest. In addition to China and India, where much tree planting is of native species in restocking, forest-rich countries such as Sweden and Finland, which rely heavily on planting to achieve regeneration, feature strongly (Appendix).

Discriminating between productive and protective functions for the planted forests component of semi-natural forests shows considerable variation between countries and regions (Table 4.1). The greater proportion, a little over 70%, is classified as productive, but a still sizeable area of planting is primarily for protective purposes of erosion control, habitat preservation, amenity and related non-industrial purposes.

4.2.2 Plantations

The Global Forest Resources Assessment 2005 (FAO, 2006a) obtained information from 229 countries. The area estimates for plantations came from responses by 150 countries reporting at least some productive plantations, but only 95 countries allocated some of their plantations to the 'protective' category. Several countries responded that data were unavailable. At least two countries, Germany and Canada, where there are believed to be plantations established according to the FRA definition, had insufficient data – Germany responding that no national plantations existed, and Canada that insufficient information was available. Furthermore, the data for some other countries which did respond are not always reliable due to the absence of recent inventories. As mentioned earlier, desk studies were undertaken for countries unable to respond to the questionnaire.

Table 4.2. Plantation areas allocated to productive and protective functions ('000 ha).

	Plantation 1990		Plantation 2000		Plantation 2005	
Regions	Productive	Protective	Productive	Protective	Productive	Protective
Africa	10,163	2,083	10,581	2,283	10,876	2,462
Asia	28,925	17,666	36,206	19,459	44,414	20,474
Europe	17,942	4,588	20,997	5,591	21,651	6,027
North and Central America	10,595	187	16,711	1,227	17,653	1,190
Oceania	2,447	1	3,477	14	3,833	32
South America	9,094	39	11,383	54	12,132	57
World	*79,165*	*24,562*	*99,356*	*28,628*	*110,560*	*30,259*
	103,727		127,984		140,819	

Table 4.2, from FRA 2005 with some minor additions, reproduces the same format as in Table 4.1 to show plantation areas by region and function for 1990, 2000 and 2005.

Table 4.2 shows the dramatic increase in areas of forest plantation reported in the 15 years from 1990. The largest resource is in Asia, where China, which has a larger area of plantations (31.7 million ha) than any other country, and to a lesser extent India, Indonesia, Japan, Thailand, Vietnam and Turkey, dominate. In Europe, the Russian federation accounts for the largest area of plantations by far (17 million ha), though several countries have well in excess of 1 million ha – France, Spain, Portugal, Ukraine and the UK. In North America, the USA, as was noted in Chapter 2, has the great bulk of the resource (17 million ha). Other countries reporting sizeable areas include Argentina, Australia, Brazil, Chile, Iran, Malaysia, New Zealand, Pakistan, South Africa and Sudan.

The relative proportions of plantations allocated to productive and protective functions are as variable as that noted earlier for the planted forests component of semi-natural forests. The proportion in the productive category – over three-quarters of all plantations – is greater, which is to be expected in view of the investment and intensity of management plantations normally entail. Perhaps what is unexpected is the high level of protective plantations reported by some countries, namely Japan (100%), Mexico (93%) and India (67%). However, this in part may reflect a question of definition and the somewhat artificial distinction between productive and protective: as has been reiterated more than once, all planted forests can serve several functions at the same time – be multi-purpose. It's just a question of the priority or objective accorded.

4.2.3 Total area of planted forest

Table 4.3 simply combines Tables 4.1 and 4.2 and shows that the planted forests subset amounts to 271 million ha in 2005, or remarkably nearly 7% of all forest

Table 4.3. Total planted forests by productive and protective functions ('000 ha).

Regions	Total planted 1990		Total planted 2000		Total planted 2005	
	Productive	Protective	Productive	Protective	Productive	Protective
Africa	11,207	2,577	11,585	2,787	11,838	3,000
Asia	64,952	35,943	74,028	40,793	86,172	45,812
Europe	54,594	13,806	60,817	15,511	63,014	16,106
North and Central America	14,571	187	24,858	1,227	27,859	1,190
Oceania	2,447	1	3,477	14	3,833	32
South America	9,119	39	11,408	54	12,158	57
World	*156,890*	*52,553*	*186,172*	*60,384*	*204,874*	*66,197*
	209,443		246,556		271,071	

in the world. Table 4.3 also shows that the total area is increasing at nearly 2% per year.

Equally remarkable is that the area of 'plantations' (Table 4.2) only represents about half of the total, yet it is 'plantations' that have attracted most of the attention in earlier resource assessments. In view of the similarities in management between the planted forests component of semi-natural forests and forest plantations generally, the expectation must be that a much higher proportion of wood production has come from this subset than hitherto realized. Perhaps the history of planting over the last 200 years should have made this obvious. Nevertheless, to discover just what the contribution is from the whole planted forests subset is the purpose of the next section.

4.3 Composition and Characteristics of Planted Forests

To grasp fully the role planted forests are playing, FAO undertook a more detailed analysis as part of its Global Planted Forest Thematic Study (FAO, 2006c). This second level of in-depth investigation is based on data and input from 36 countries (see Fig. 4.1), though it should be noted that not all were able to respond with information for every question in the survey. In view of this more limited database – the actual number of countries and varying levels of response – it is only possible to generalize rather than extrapolate with great confidence. That said, the data are the best so far assembled and highlights are summarized here as the foundation for Chapter 5. They are considered under the two broad functions of productive and protective, and in each comment is made on species, growth rates, age classes and rotations, ownership and end uses.

4.3.1 Planted forests for production

The rate of establishment of new planted forests for production is increasing slightly, from 1.7% yearly in 1990–2000 to 1.9% yearly in 2000–2005. This disguises a contrast between Asia, with 42% of the world's planted forests for production, where new planting is adding 3% per year to area since 2000, and Europe, with the next largest area (31%), where expansion has slowed considerably to just 0.7% yearly. The relative proportion of planted semi-natural forests to plantations is almost 50:50, though the trend is clearly towards forest plantations.

Species used for production

A small range of species is used for production in both tropical/subtropical and temperate regions. Several subregions in the tropics and subtropics plant the same species, of which the following are the main ones: *Acacia mangium*, *A. nilotica*, several eucalypts (especially *E. camaldulensis*, *E. grandis* and in cooler parts *E. globulus*), *Gmelina arborea*, *Tectona grandis* and large areas of pines (notably *Pinus caribaea*, *P. elliottii* and *P. patula*). For the tropics it should be noted that some species that formerly provided non-forest products, such as rubber (*Hevea brasiliensis*) are being used as wood or fibre for industrial products. Also in the cooler subtropics of China can be found the world's most planted tree species in terms of area, Chinese fir (*Cunninghamia lanceolata*).

In Mediterranean and temperate regions *E. globulus*, poplars (*Populus* spp.) and pines (*Pinus* spp.) are important. For the latter, most notably *Pinus radiata* in the southern hemisphere, but elsewhere *P. massoniana*, *P. nigra*, *P. pinaster*, *P. sylvestris* and *P. taeda*. In cooler temperate and boreal regions, as well as *Pinus* other conifer genera of *Abies*, *Larix*, *Picea* and *Pseudotsuga* are also important.

The definition of the planted component of semi-natural forests is that they should be predominantly of native species. Examples include *T. grandis* in India and Thailand, *C. lanceolata* in China, *Araucaria* spp. in Oceania and loblolly pine (*P. taeda*) and Douglas fir (*Pseudotsuga menziesii*) in the USA. Also in this category are plantings of *Acacia senegal* and *A. seyal* in North Africa, *Khaya* in West Africa, *Cordia* and *Prosopis* in central and South America, and oak (*Quercus*) and beech (*Fagus*) in Europe and West Asia, as well as species already mentioned such as *P. sylvestris* in Europe and Asia.

However, forest plantations have been dominated by the use of exotic species because of gains from careful matching of species with sites and relative freedom from pest and disease problems, which together have resulted in high growth rates (Zobel *et al.*, 1988). Of the several thousand tree species in the world, only about 30 have been widely planted. And of these, with the exceptions of teak (*T. grandis*) and Chinese fir (*C. lanceolata*), most are from just four genera, namely, *Acacia*, *Eucalyptus*, *Pinus* and *Populus*. This restriction offers both opportunity and constraint: opportunity for collaboration, for greater in-depth understanding of a species and its potential; and constraint from ignoring the potential of many, many other species. It is recognized that numerous species and provenance trials have evaluated thousands of taxa, but the facts remain that success with planting valuable tropical hardwoods has been meagre and throughout the world many native species are shunned because of slower growth and greater costs of cultivation and silviculture.

Growth rates, rotations and age classes

Information was sought on the growth rates of the main species established for productive purposes. Data, even from the 36 countries surveyed, were incomplete and rather than citing specific examples it is more helpful to indicate orders of magnitude. What is clear is that growth rates from the planted component of semi-natural forests are lower than those from forest plantations owing to the use of native species and associated longer rotations.

Table 4.4 summarizes productivity data, drawing on the FAO Global Planted Forest Thematic Study (FAO, 2006c) and Evans and Turnbull (2004).

Age-class distributions from countries surveyed showed a reverse-J curve, with most of the area of planted forest in the younger age classes, established since the 1980s, as would be expected both from the upsurge in planting in recent decades and from the rotation age data for many of the planted forests concerned. Younger age classes predominate in forest plantation compared with the planted component of semi-natural forests which have a proportionally greater area over the age of about 40 years. This confirms the shift towards the establishment of more forest plantations, which was particularly apparent in data from East Asia.

Ownership

Ownership of forest and other wooded land was reported for all forests in FRA 2005 (FAO, 2006a) but not specifically for the planted forest subset. The FAO thematic study on planted forest shows that recent trends in decentralization and devolution arising from privatization (with or without increased participation) are reflected in the ownership of planted forests for production. The area in public ownership in 1990 formed 70% of the total area, but by 2005 had fallen to 50%. Corporate ownership (generally large scale) has remained steady at about 18% of the total area. The change has been that of smallholder ownership, rising from 12% in 1990 to 32% in 2005. This is dominated by a dramatic increase in this

Table 4.4. Growth rate and rotation data by species and regions for planted forest.

Species group	Region	MAI range (m^3/ha/year)	Rotation (years)
Broadleaves	Wet tropics	10–45 [15][a]	5–15
	Dry tropics	5–15 [5]	10–30
Pines	Tropics/subtropics	10–25 [12]	10–30
Tectona grandis	Dry tropics	4–10 [5]	25–70
Cunninghamia lanceolata	Central China	3–12 [5]	15–30
Pinus radiata	Mediterranean regions	10–30 [15]	15–35
Populus spp.		4–20 [10]	10–30
Abies, Larix, Picea and *Pinus*	Temperate	2–16 [8]	30–150
Fagus and *Quercus*	Temperate	2–10 [4]	80–200
Boreal conifers	Boreal forest	1–10 [3]	60–150

[a]Figures in square brackets indicate a conservative average for the category.

category in China but also from greater interest in private ownership of usually small woodlands and forests in, for example, both the UK and USA. Although there are large differences between regions and between countries, the trend from public ownership to private ownership, especially by smallholders, is clear. Further encouragement of this trend through facilitating policies seems likely, but planning wood supplies for industry could become more difficult and there is the danger of smallholders moving to another form of land use that may be more profitable.

End use of planted forests

Twenty-four of the 36 surveyed countries, representing 80% of sampled planted forest, reported end-use purpose (see Table 4.5).

Nearly half of all planted forests for production are reported as managed for saw-log or veneer-log production, which accords with forest products statistics of production from all forests (FAO, 2005), though the proportion has been decreasing since 1990. Also since 1990, the area planted for pulpwood/fibre has more than doubled and now amounts to nearly one-fifth of the total. The increase in the amount of non-wood products shown in Table 4.5 is probably due to inclusion of rubberwood. The small changes in planted forests for bioenergy don't yet reflect the gathering upsurge in biomass plantations as a response to climate change.

4.3.2 Planted forests for protection

The extent of planted forest established for protective purposes has gradually increased since 1990 to over 66 million ha in 2005 (Table 4.3). Of this total rather more than half is in the planted component of semi-natural forests. As with planted forest established for production, and as commented on earlier, the largest area of planted forest for protective purposes (Fig. 4.3) is in Asia, and accounts for two-thirds of the total.

Table 4.5. Planted forests, industrial end use ('000 ha and % of annual total).

End use	1990	2000	2005
Pulpwood/fibre	11,783	24,852	26,741
	11.1%	18.8%	18.4%
Sawlogs	57,194	62,855	67,099
	54.1%	47.7%	46.5%
Bioenergy	7,115	7,364	8,497
	6.7%	5.6%	5.6%
Non-wood products	10,530	17,319	23,359
	10.0%	13.1%	16.1%
Unspecified	19,127	19,526	19,477
	18.1%	14.8%	13.4%
World	105,749	131,916	145,173

Fig. 4.3. *Pinus halepensis* planted on gradoni (narrow terraces) to help control soil erosion in Jordan. (Source: FAO.)

The more detailed analysis below is weaker than that just concluded for planted forests for production owing to a small database from fewer responses from the countries surveyed.

Species

While many species planted for production purposes are also used for protective planting – after all a species that grows well on a site is usually a good choice whatever the objective – native species are clearly preferred even for forest plantations. Examples include *Cryptomeria japonica* and *Larix kaempferi* in Japan, *Acacia senegal* and *Pinus halepensis* in North Africa, and *P. sylvestris*, *Picea abies* and *Larix decidua* in Europe.

In passing it should be noted that some species serve poorly for protection, including teak (*T. grandis*) and many *Eucalyptus* spp., owing to the type and quantity of their litter and suppression of understorey.

Growth rates, rotations and age classes

It is difficult to attribute yield data to protective planted forest because that is not its principal objective. But, equally, many such forests will yield some produce as a secondary function, and many will be deemed 'multi-purpose' while accepting that protection is the more important objective. For estimates of biomass, carbon sequestration and forecasts of wood production, it is probably acceptable to use mean growth rates equal to half of the averages shown in square brackets in Table 4.4 and rotation lengths at the older end of the age range, i.e. assume slow growth and long rotations and hence much reduced yield.

The global distribution of the age classes of planted forests for protection was close to 'normal', that is an even spread of areas between age classes – at least up to the age of 50 years.

Ownership

From the limited survey carried out, most (70%) planted forests with protective function were publicly owned in 2005, with the balance in corporate ownership. Both in China and to a much smaller extent in Europe, there has been a trend towards smallholder ownership, although less pronounced than with planted forests for production.

Most publicly owned planted forests for protection are located in East Asia, with large areas in China and Japan, and in Europe, where the bulk is in the Russian Federation and to a lesser extent in countries such as Poland. Many publicly owned protective forests are in the semi-natural category, reflecting reservation by the state of existing natural forest for protection purposes but with reliance on planting to ensure satisfactory re-stocking and continuance of cover. However, in Russia many forest plantations are allocated to protective functions, as is also the case in other countries for corporately owned planted forests, indicating that corporate owners set aside parts of their plantation estate for protection purposes such as riparian zones, buffers next to enclaves of native woodland etc., perhaps stimulated by forest certification pressures.

End uses of protective forest

The limited survey results indicate that, not surprisingly, the protective end use is dominant, namely, protection of soil, water and biological diversity – including the harvesting of non-wood forest products on a small scale. Such forests also commonly fulfil amenity and recreation functions, which for some countries, such as Japan and Poland, are the dominant purpose.

It is recognized, of course, that many planted forests with a protective function are truly 'multi-purpose' and the constraint of classifying by end use is artificial to the point of being misleading. Nevertheless, the picture emerging shows a planted forest resource with many facets and providing many products and services – wood and non-wood – where some purposes are more important than others and hence their *raison d'être*.

4.4 Impact of Planted Forests

The impact of the planted forests subset of the world's forest resources, and on the global forestry sector, leads to several conclusions.

- The estimates of area, 271 million ha amounting to 7% of the world's forest, point to a far more significant resource than hitherto recognized.
- Planted forests are continuing to expand, pointing to an even greater role in the future.

- Distinguishing between planted forests as a component of semi-natural forests and those traditionally labelled forest plantations has revealed better the extent and impact of plantation silviculture as a contribution to forest management and practice globally.
- Policy making and planning, as well as the allocation of funds for protection (fire, pests and disease), research and maintenance, will need to take into account the increase in planted forests in general and the trend towards forest plantations.
- Data on growth rates, rotations and end uses suggest that planted forests contribute massively to industrial wood and fibre supply, which, in the past, may have been underestimated. But just how large the contribution might be is the subject of Chapter 4.
- Distinguishing between functions of planted forest – production and protection – reveals the substantial investment in environmental protection to obtain services from forest cover (the usually benign woodland influences) to conserve soil, water and wildlife, to protect slopes and provide shelter, and to give enjoyment for recreation and amenity at local and landscape level.
- The impact of the production planted forests may be nearly as great on environmental protection as protective planted forests may provide in economic and social benefit terms to, especially, rural people. Achievement of these multiple objectives will depend on management that considers all economic, social and environmental aspects.
- The impact of large volumes of certain species and size classes, which have within the past 25 years been widely established for productive purposes, on markets in the near future must be considered by planners – and technologists who may, for example, have to develop ways of utilizing large quantities of small-dimension logs.
- The impact on wood supply as well as on the provision of environmental services of the migration into private ownership of planted forests, including to smallholders, is a critical shift raising some uncertainties about continuity of supply which may need addressing through policy-related measures.
- Finally, it is recognized that data are incomplete. Some countries with known major planted forests resources have not adequately contributed to these data, and some geographical regions such as the dry subtropics are poorly represented. Further research will refine these conclusions.

4.4.1 The future fibre basket

The above list indicates many striking impacts that the new planted forests data point to. Perhaps, most striking of all is the past underestimation of the role planted forests played in the supply of forest products. In one generation a sea change has occurred. The optimism about 'man-made' forest and the speculation concerning planting programmes 40 years ago at the Symposium on Man-made Forests and their Industrial Importance (FAO, 1967) have, it can be argued, not only been realized but exceeded (Fig. 4.4).

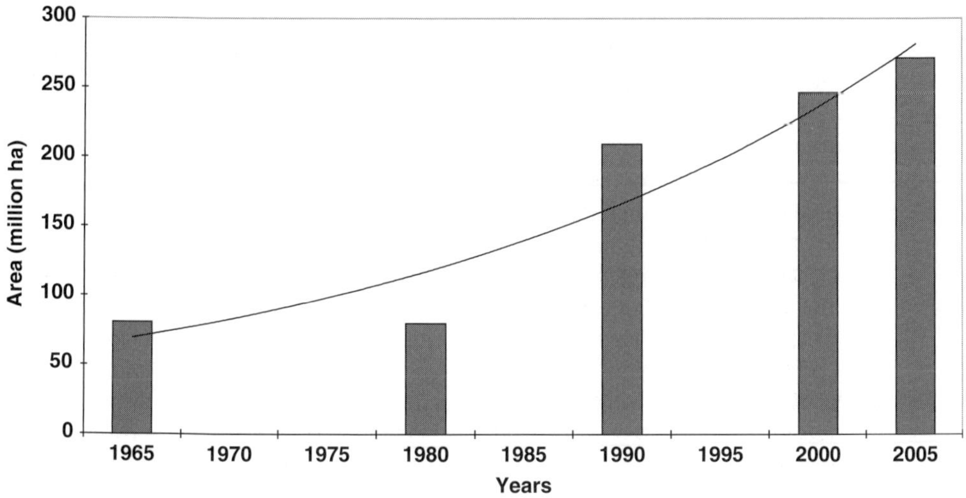

Fig. 4.4. Estimated areas of planted forests ('000 ha), 1965–2005. Note: the early data are both incomplete and suffer from changing definitions.

While the wider definition of planted forest explains part of the acceleration in area terms (Fig. 4.4) – or rather paints a truer picture of the contribution they actually play – there is no doubt that planted forests are on the point of exceeding the production coming from natural forest formations (primary, modified, and semi-natural forest with assisted natural regeneration in Table 3.2).

The evidence is straightforward. Table 4.5 indicates that around 175 million ha of planted forest (the 145 million ha in Table 4.5 is believed to represent 80% of planted forests for production) have defined production objectives. Using a very conservative MAI of 5 m^3/ha/year [the same figure hypothesized in FAO (1967)] for all these planted forests wherever they are, their potential annual yield is still 875 million m^3, or already equal to half current world consumption of industrial forest products. Bearing in mind that: (i) many planted forests for protection will also yield some worthwhile produce; and that (ii) extensive forest plantations in the tropics, subtropics and several southern hemisphere countries far exceed an average MAI of 5 m^3/ha/year, it is reasonable to argue that planted forests already supply in the region of 1 billion m^3/year of wood.

This very rough and ready but remarkable estimate of yield from planted forests brings a paradigm shift. Just 7% of the world's forest – the planted forest component – can potentially produce two-thirds of global industrial roundwood. With the investment in genetic tree improvement, strategies to focus planting on best-adapted sites and other silvicultural advances, a further lift in productivity per hectare is guaranteed. Overall fibre supply from planted forest will increase and eliminate any lingering spectre of wood shortage globally, if not always locally. Planted forests are fast becoming the world's fibre basket and will complete the domestication of forest production and so reach what agriculture achieved centuries ago.

The significance of this emerging evidence is far too important to leave as crude generalizations. Consequently a detailed study was undertaken to attempt to reach global and regional forecasts of output of industrial forest products from all countries with significant planted forest resources: this is the subject of the next chapter (Chapter 5). The far-reaching implications concerning the carbon cycle, bioenergy, and all the related social, institutional, ecological and environmental issues, are examined in Chapters 6–8.

Note

1 For definition of 'assisted' see Endnote 2, Chapter 3 and Table 3.2.

5 Wood from Planted Forests: Global Outlook to 2030

J.B. Carle and L.P.B. Holmgren

5.1 Introduction

The broadening of definition to include the planted semi-natural forests not previously reported doubles the area in the planted forests subgroup and has a substantial impact on the yields of forest products and social and environmental services. According to FAO (2005), there were 140 million ha of forest plantations globally, of which 78% were for productive purposes. According to the Global Planted Forest Thematic Study (FAO, 2006c), the global planted forest area was estimated at 271 million ha, of which 76% was for productive purposes. This chapter explores the impact of this enlargement of definition and considers alternative global outlooks for the provision of wood from planted forests from 2005 to 2030 (Carle and Holmgren, 2008).

5.1.1 Recent outlook studies

The global outlook for plantations (ABARE and Pöyry, 1999) and the global outlook for future wood supply from forest plantations (FAO, 2000) provide the most comprehensive and recent studies. Both were based upon FAO's Forest Resources Assessment 1990 dataset, updated to 1995 in 1997, and used the prevailing forest plantation definitions in FAO (1993).

The ABARE study estimated that although the productive forest plantation area was 116 million ha or about 3% of global forest area in 2000, forest plantations were estimated to produce 35% of global wood supply in 2000, 44% in 2020 and 46% in 2040.

The FAO (2000) outlook study explored three scenarios of future forest plantation expansion and three different extrapolations for industrial roundwood consumption to 2050. In 1995 it was estimated that there were 124 million ha of forest plantations (3.5% of forest area) that yielded over 22% of industrial roundwood production, and this would be, by 2010, between 31 and 34%, by 2020 up to 46%, and by 2050 up to 64%, depending upon the scenario of forest plantation production and industrial roundwood consumption.

5.1.2 Policy context

Planted forests are recognized as a valuable land use to realize the values and principles of the Millennium Development Goals, particularly to: eradicate extreme poverty and hunger (goal 1); ensure environmental sustainability (goal 7); and develop global partnerships for development (goal 8). Despite being just 2% of global land use, planted forests play an important role in the provision of a wide range of goods – roundwood (industrial and subsistence), fibre, bioenergy and non-wood forest products – and social and environmental services (conservation, protection of soil and water, rehabilitation of degraded lands, combating desertification, carbon sinks, recreation, diversification of urban and rural landscapes and employment). Responsible management of planted forests can reduce the pressure on the range of goods and services provided by native forests and enhance the livelihoods of local communities, including indigenous peoples.

In recent years a diverse modern forest industries sector has been encouraged to adapt to the use of the 'new wood' from planted forests. The range of industrial products from planted forests include: lumber, plywood and veneer, reconstituted panels (MDF, OSB, chipboard etc.), modular components (laminated products, moulding, framing, floorings etc.), pulp and paper and increasingly bioenergy. Scientific research and development, particularly in genetic improvement and forest industries processing, have revolutionized the productivities and the end-use options for planted forests. Application of biotechnology has substantially improved site-species matching, growth, yields and financial benefits for planted forest investors, particularly in fast-growing, short-rotation crops. The development of forest industries technology has resulted in increasing end-use options for raw materials from planted forests, improved efficiencies and reduced wood industries costs (Barbour, 2004; Bowyer, 2004; Sedjo, 2004; Youngs, 2004).

Industrial roundwood from planted forests is being recognized as a renewable resource and an energy efficient and environmentally friendly raw material for construction when compared with alternative products such as steel, aluminium, concrete and plastic (Bowyer, 2004). Planted forests can make significant positive contributions to rural economies through primary and secondary industry development, employment and development of rural infrastructure. Trees are increasingly being planted to support agricultural production systems, community livelihoods, poverty alleviation and food security.

5.1.3 Outlook objectives

The study reported here seeks to estimate the industrial roundwood from planted forests globally. Whilst recognizing the important social and environmental services from planted forests, which are explored in Chapter 6, these aspects are not addressed here.

Table 5.1. Summary of planted forest area in the 61 studied countries by region and major species groups at 2005.

	Softwoods		Hardwoods			
	Pinus spp.	Other	*Acacia* spp.	*Eucalyptus* spp.	Other	Total
Region	Million ha	Million ha	Million ha	Million ha	Million ha	Million ha
Africa	1.2	0.5	5.2	1.2	1.4	9
Asia	18.9	15.3	3.8	7.6	79.2	125
North, Central and Eastern Europe	26.4	36.0	–	–	12.1	74
Southern Europe	0.0	4.6	–	0.0	4.7	9
North and Central America	18.9	7.2	–	–	1.7	28
South America	5.1	0.3	0.2	4.5	0.9	11
Oceania	2.7	0.2	0.0	0.5	0.2	4
Total	**73**	**64**	**9**	**14**	**100**	**261**

5.2 Building the Model

5.2.1 Countries surveyed

The baseline data come from the status of planted forests in the 61 countries noted in Chapter 4 and are believed to represent about 95% of the estimated global planted forest area of 271 million ha in 2005 (FAO, 2006c). As noted, the survey requested in-depth information about planted forests in each country, including species distribution, ownership, end use, rotation lengths, mean annual increment (MAI) and age-class distribution. Of the 61 countries, 36 responded to a formal information request, and 25 were subject to a desk study (FAO, 2006c). This outlook is limited to these 61 countries and thus will give somewhat conservative results for global planted forests. A summary of the initial state is presented in Table 5.1.

5.2.2 Scenarios explored

Three scenarios are examined to take into consideration potential changes in the planted forest area (mainly through new plantings) as well as opportunities for increased productivity resulting from more efficient management practices, new technology and genetic improvements (Table 5.2).

Table 5.2. Description of the three scenarios applied in the outlook model.

Scenario 1 – Pessimistic scenario
Area changes are assumed to be half of the predicted ones for Scenario 2, and there are no productivity increases. This represents a scenario where the current increase of planted forest area will slow down.

Scenario 2 – Business as usual
Area changes have been predicted based on past trends and are assumed to continue at the same rate until 2030. However, there are no productivity increases in this scenario.

Scenario 3 – Higher productivity
Area changes have been predicted as in Scenario 2. In addition, an annual productivity increase has been applied for those management schemes where genetic, management or technological improvements are expected. As an example, a productivity increase of 2% annually equals an accumulated productivity increase of 64% for the 25-year period (2005–2030).

5.2.3 Management alternatives

The unit of analysis in the outlook is a 'management scheme', defined by country, species/species group, purpose (protective or productive) and characteristic (plantation or semi-natural forest) of the planted forest subset (FAO, 2006c). Parameters applied in the outlook model for each management scheme are listed in Table 5.3, together with one example management scheme: *Picea sitchensis* in Ireland. In total, over 660 management schemes were identified for the 61 countries and applied in the modelling. Input data missing from the country survey and data for area efficiency and productivity changes were filled through expert estimates. All management scheme input data are given in Carle and Holmgren (2008). A summary of the management scheme inputs is shown in Table 5.4 and Fig. 5.1.

5.2.4 Modelling

A deterministic model was developed using Microsoft Excel to predict future production of wood in each management scheme, for each of the five wood end-use categories, following the process in Fig. 5.2. The model was run for every management scheme and for each of the three scenarios for the period 2005–2030. To derive longer term projections at a more general level, the rotation length distribution in Table 5.4 was used to create a simplified set of 11 management schemes, for which the model was run for each of the three scenarios for the period 2005–2105.

5.3 Results

5.3.1 Area trends

Model results indicate that the area of planted forests will increase in all scenarios, as shown in Table 5.5. From an initial area of 261 million ha, the area

Table 5.3. Model input parameters for each management scheme.

Parameter	Unit	Comment	Example: Ireland, *Picea sitchensis* – forest plantation; productive purpose
Area	ha	Total extent of the management scheme.	301,080 ha
Age-class distribution	%	Distribution of the area across 12 age classes. The sum of the 12 proportions to be 100.	1–5: 10% 6–10: 10% 11–20: 23% 21–30: 24% 31–40: 20% 41–50: 9% 51–60: 2%
Rotation length	Years	Average rotation length across the management scheme.	50 years
Mean annual increment (MAI)	m^3/ha/year	Average growth in stem volume on bark as average over rotation cycle and across the management scheme.	18 m^3/ha/year
Area efficiency	%	An estimate of the relative performance (max 100%) of the management scheme, taking into account (a) reductions of overall area related to infrastructure or unsuccessful stand establishments, (b) reduced productivity due to stand health issues or suboptimal management practices, (c) influence of other management objectives, particularly related to protective functions, on the wood volume production.	90%
Volume end use for: Fuel/bioenergy Pulp/fibre Wood products Unspecified Harvest losses	%	Distribution of expected end use of stem wood into five categories as listed in the first column. The sum of the five proportions should be 100.	Fuel/bioenergy: 5% Pulp/fibre: 30% Wood products: 60% Unspecified: 0% Harvest losses: 5%
Annual area change	%	The annual increase in area (net new establishments). The increase is applied in relation to the initial area throughout the studied time period, i.e. as a linear development. This parameter varies between the applied scenarios.	Scenario 1: 1.5% Scenarios 2 and 3: 3%
Annual productivity change	%	The annual increase in overall productivity, representing improved area efficiency (see above), better management practices, higher technology efficiency and genetic improvements. The increase is applied to the previous year throughout the studied time period, i.e. as an exponential development. This parameter was applied only in scenario 3.	Scenarios 1 and 2: 0% Scenario 3: 1%

Table 5.4. Summary of planted forest area and model input parameters for different rotation lengths at 2005 for the 666 management schemes identified.

Rotation length (years)	Area (million ha)	MAI (m^3/ha/year)	Production potential Area*MAI (million m^3/year)	Management schemes included (n)	Average area efficiency (%)	Average area expansion (scenarios 2+3) (%/year)	Average productivity increase (scenario 3) (%/year)	Proportion young stands[a] (%)	Proportion aged stands[a] (%)	Proportion over-aged stands[a] (%)
<10	13	23	288	43	76	1.46	1.89	53	30	17
11–20	25	10	240	60	63	2.85	1.41	43	30	27
21–30	64	10	615	90	72	0.84	0.52	55	36	8
31–40	38	7	251	71	58	2.40	0.34	58	25	17
41–50	16	8	129	48	67	1.11	0.55	44	39	17
51–60	23	8	187	60	69	1.44	0.74	63	31	6
61–70	39	7	278	60	77	0.54	0.13	57	39	4
71–80	15	7	100	44	70	0.52	0.28	80	20	0
81–90	11	5	53	36	93	1.81	0.80	81	17	2
91–100	2	6	15	31	68	0.74	0.12	78	20	2
101+	14	7	91	123	62	0.00	0.01	57	40	3
Total	**261**	**9**	**2246**	**666**	**70**	**1.29**	**0.58**	**58**	**31**	**11**

[a] Young stands defined as aged <0.5 * rotation length, aged stands defined as aged 0.5–1 * rotation length, over-aged stands defined as aged > rotation length.

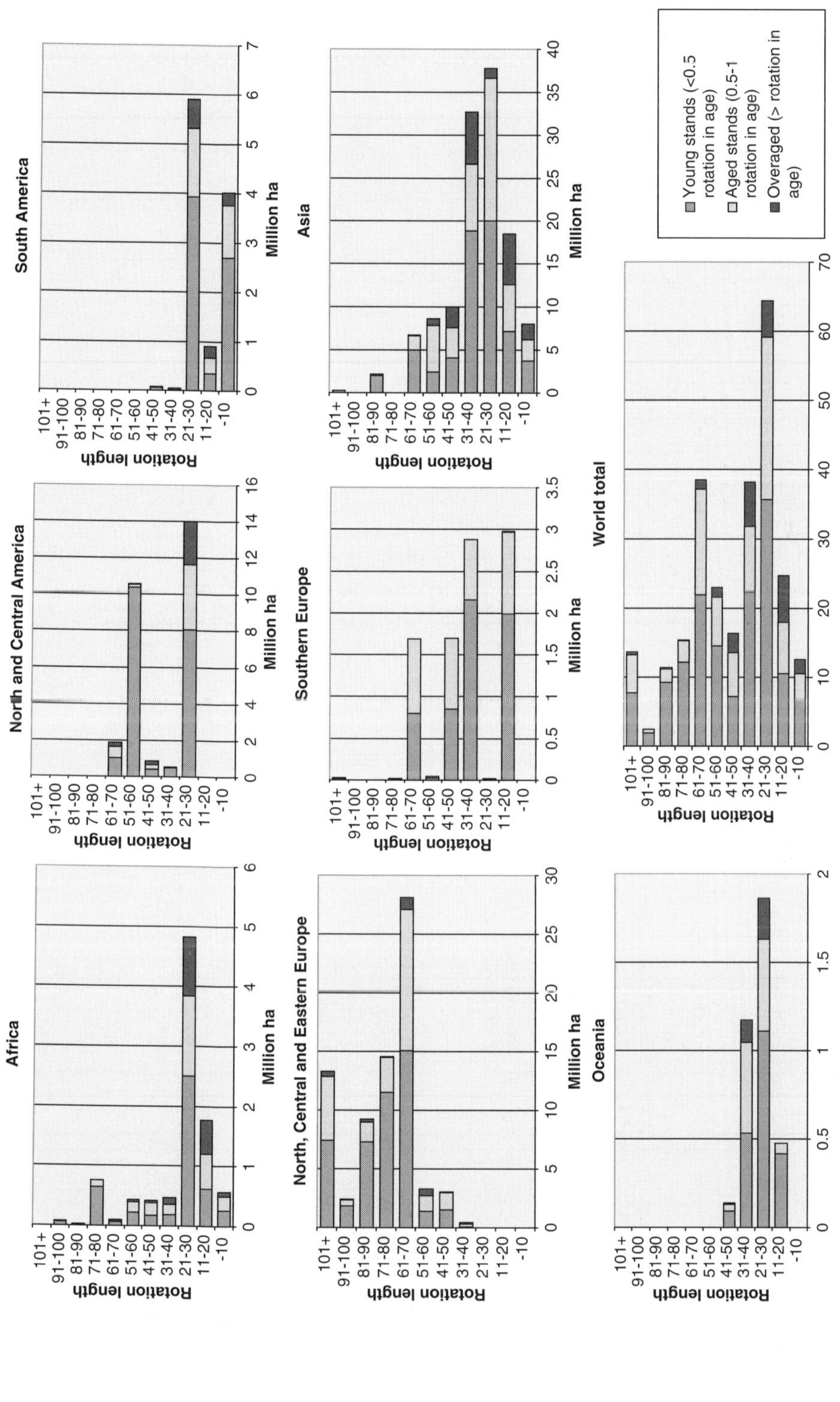

Fig. 5.1. Distributions of rotation lengths (years) and maturity of stands in relation to rotation length by region.

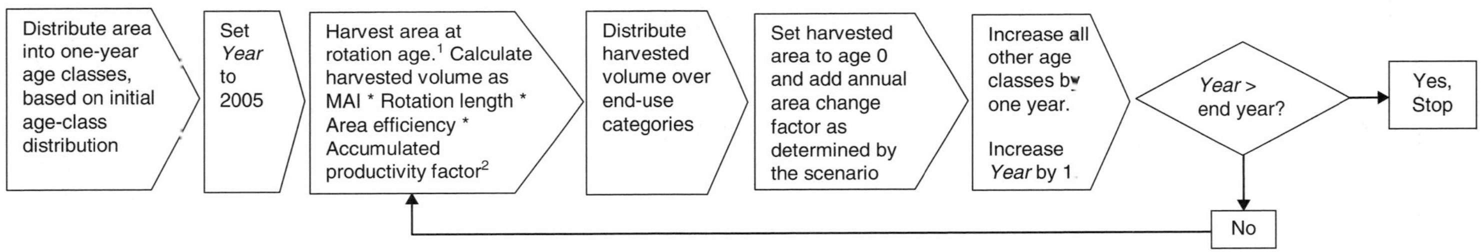

[1] Initially overaged stands are harvested gradually over a 10-year period.
[2] Productivity factor is only applied for Scenario 3 (see Tables 5.2 and 5.3).

Fig. 5.2. Outlook model process applied to each identified management scheme for each scenario.

Table 5.5. Area of planted forests by region and major species groups at 2005 and 2030 for the three scenarios (millions ha).

Region	*Acacia*	*Eucalyptus*	*Pinus*	Other softwood	Other hardwood	Total
2005						
Africa	5.2	1.2	1.2	0.5	1.4	9.4
Asia	3.8	7.6	18.9	15.3	79.2	124.8
North, Central and Eastern Europe			26.4	36.0	12.1	74.5
Southern Europe				4.6	4.7	9.3
North and Central America			18.9	7.2	1.7	27.8
South America	0.2	4.5	5.1	0.3	0.9	10.9
Oceania		0.5	2.7	0.2	0.2	3.6
Total	**9.1**	**13.8**	**73.2**	**64.0**	**100.3**	***260.5***
2030, Scenario 1						
Africa	4.7	1.2	1.4	0.5	1.6	9.4
Asia	4.6	10.6	23.3	16.9	92.8	148.3
North, Central and Eastern Europe			28.8	38.3	12.5	79.6
Southern Europe				7.5	7.6	15.0
North and Central America			21.9	9.8	2.0	33.7
South America	0.2	5.2	6.0	0.3	1.0	12.7
Oceania		0.7	2.8	0.2	0.3	3.9
Total	**9.5**	**17.7**	**84.2**	**73.5**	**117.8**	***302.7***
2030, Scenarios 2 and 3						
Africa	4.2	1.2	1.6	0.5	1.8	9.4
Asia	5.4	13.6	27.6	18.5	106.4	171.7
North, Central and Eastern Europe			31.3	40.6	13.0	84.9
Southern Europe				10.3	10.4	20.8
North and Central America			25.0	12.5	2.4	39.8
South America	0.2	5.7	6.5	0.4	1.1	13.9
Oceania		0.8	2.9	0.2	0.3	4.2
Total	**9.9**	**21.4**	**94.9**	**83.0**	**135.5**	***344.6***

increase in scenario 1 will be 16% to 303 million ha by 2030, and in scenarios 2 and 3 an increase of 32% to 345 million ha. Among regions, the highest absolute increase will be in Asia and the highest relative increase in Southern Europe. Among species groups, the highest absolute increase will be *Pinus* forests.

5.3.2 Wood volume trends

The model's estimates for wood production by species groups and regions for the period 2005–2030 are given in Fig. 5.3 and Table 5.6. The total volume produced increases from 1.4 billion m^3 in 2005 to 1.6, 1.7 and 2.1 billion m^3 respectively in the three scenarios. Most of the variation between scenarios arises in Asia and South America, where the higher productivity scenario gives a considerable increase in wood production. The differences between scenarios 1 and 2 are small, primarily because the additional planted area in scenario 2 may not generate much wood within the study period.

Figure 5.3 also illustrates that South America and Asia have a more dynamic future, compared with other regions, and, for scenario 3, that the volume increases will mainly be in *Eucalyptus* and other hardwood species.

Table 5.6 illustrates that the proportion of wood for industrial use (the sum of the end-use categories pulp/fibre and wood products) is about 85% of all wood from planted forests. The total volume of wood for industrial use increases from 1.2 billion m^3/year in 2005 to 1.9 billion m^3/year in 2030 according to scenario 3.

5.3.3 Long-term projections of wood volume

Figure 5.4 shows the impact of projecting the model's scenarios for up to 100 years. The continued linear increases of wood volumes in scenarios 1 and 2 are confirmed, with volume production from scenario 1 reaching about 2.5 billion m^3/year and that from scenario 2 about 3.5 billion m^3/year. Scenario 3, which incorporates a continuing steady increase in productivity, postulates a much more rapid development of wood production to about 9 billion m^3/year in 2105.

5.4 Some Implications

While it is acknowledged that any model is only as good as the assumptions made and the quality of data used, the roundwood production forecasts from the world's planted forests confirm their dominance in future industrial wood supply. They far exceed earlier projections (e.g. FAO, 2000) because of the broadening, and it is argued more realistic, definition of what constitutes planted forest. This has many implications for the world's forests, in practice and in perception. As commented earlier, the 'domestication' of wood production has come of age.

Many factors can disturb these projections, but the possible negatives (pest and disease threats to planted forests, loss of political will to continue planting, competing land uses, environmental concerns etc.) appear more than compensated by likely positives (much expanded planting for carbon storage, increased yields per hectare from genetic and silvicultural improvements, increased

Table 5.6. Wood volume produced in planted forests by region and use at 2005 and 2030 for the three scenarios (million m^3/year).

	Fuel/ bioenergy	Pulp/ fibre	Wood products	Unspecified	Harvest losses	Total
2005						
Africa	11	9	55	6	1	82
Asia	79	141	264	6	5	495
North, Central and Eastern Europe	17	123	166	8	15	329
Southern Europe	3	26	26	0	3	58
North and Central America	7	98	24	0	7	135
South America	19	133	91	0	10	253
Oceania	1	11	31	0	4	47
Total	**136**	**540**	**659**	**21**	**44**	***1400***
2030						
Africa	10	14	57	6	2	89
Asia	83	132	311	18	6	550
North, Central and Eastern Europe	18	129	185	4	17	353
Southern Europe	5	44	45	0	5	98
North and Central America	7	106	29	0	7	149
South America	21	157	106	0	12	295
Oceania	1	12	35	0	4	53
Scenario 1, Total	**146**	**593**	**767**	**29**	**53**	***1589***
Africa	10	15	56	6	2	89
Asia	88	146	321	20	7	582
North, Central and Eastern Europe	18	129	185	4	17	353
Southern Europe	6	55	56	0	6	123
North and Central America	8	117	31	0	8	164
South America	23	173	115	0	13	323
Oceania	1	13	36	0	4	55
Scenario 2, Total	**155**	**647**	**800**	**30**	**57**	***1689***
Africa	10	22	63	6	2	103
Asia	107	204	417	22	7	756
North, Central and Eastern Europe	20	137	200	4	17	378
Southern Europe	8	67	69	0	6	150
North and Central America	10	149	38	0	8	206
South America	34	268	156	0	13	471
Oceania	2	19	55	0	4	81
Scenario 3, Total	**191**	**866**	**998**	**33**	**57**	***2145***

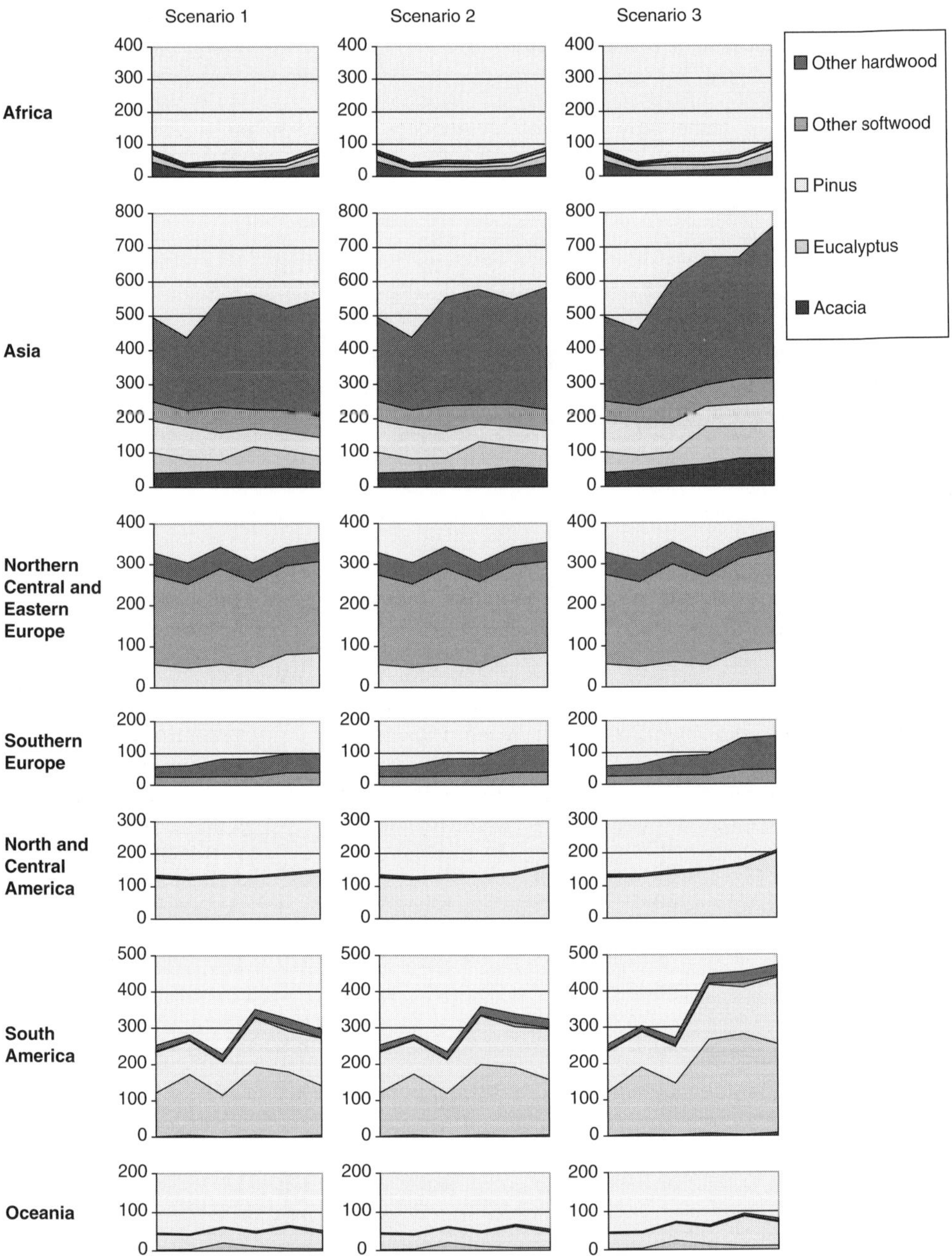

Fig. 5.3. Wood volume produced in planted forests 2005–2030 by major species group and region, for each of three scenarios (million m^3/year).

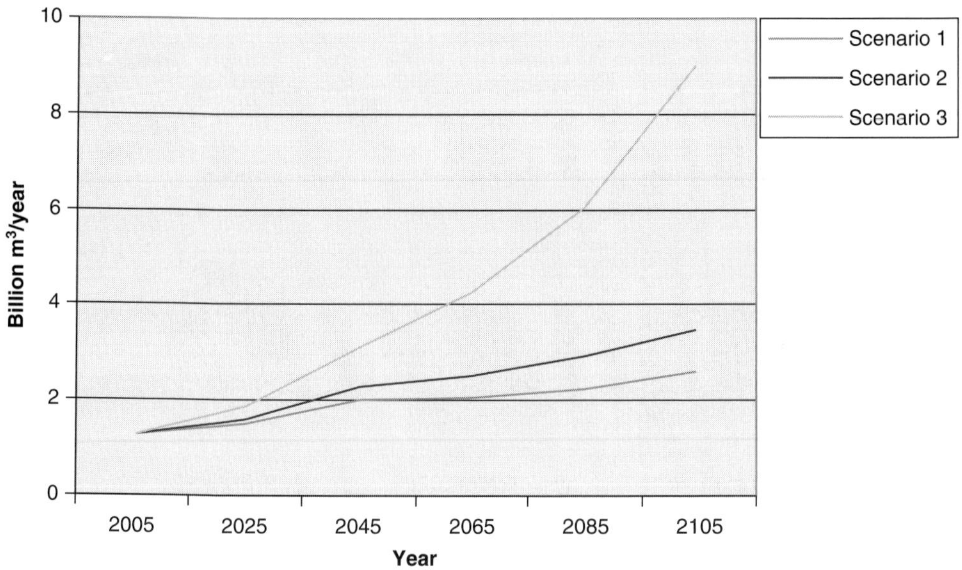

Fig. 5.4. Long-term (100-year) projection of total global wood production from planted forests for the three scenarios.

interest in planted forest as an investment, the rise of the smallholder plantings etc.) and suggest that the projections are realistic. How realistic they are, what roles planted forests will play, what policy and institutional issues may arise, and what this means for sustainability as a whole, are the subject of the next three chapters.

6 The Multiple Roles of Planted Forests

J. EVANS

6.1 Introduction

Planted trees and planted forests serve many roles and functions. This review of issues and developments seeks a positive approach in advocating and examining how best to take advantage of such management and silviculture to balance the social, cultural, environmental and economic trade-offs. For example, as Ghazoul and Evans (2004) point out, it is poor forest stewardship simply to clear native forest to provide somewhere to plant trees. Not only are adverse impacts many, but there is plenty of degraded, waste or cut-over land entirely suitable for planted forest. A second introductory point is one that is frequently made (Evans, 2004b) and made earlier in this book: planted trees and forests are able to serve several roles at once. It is harnessing trees for their wood, fibre and non-wood forest products, their many influences (environmental and social), and their interaction in the ecosystem for the good of all which are the aims. In general, planted forests do not seek to mimic the functions of native forests but are managed intensively to meet stated purposes.

6.2 Wood, Fibre and Energy Production

6.2.1 Expectations for industrial wood production

Chapters 4 and 5 clearly point to the rapid ascendancy of planted forests and, necessarily, the forest products they supply. In one generation a change has occurred. Figure 4.4 traced the expansion in area, but in a single lifetime the proportion of global production of industrial roundwood from planted forests has risen from a few per cent to potentially two-thirds (2008), with well over this figure predicted for the future, as the previous chapter indicates. While the wider definition explains part of this acceleration – according to Youngs (2004), industrial forest plantations alone supplied 35% of global roundwood in the year 2000 – there is no doubt that production is rapidly shifting away from native forest formations to planted forests, and they will become the principal source of production. This is not a new observation (e.g. Sedjo, 2004; Youngs, 2004), more a new appreciation or understanding. As noted earlier, it completes the domestication and intensification of forest production, reaching what agriculture achieved centuries ago.

The estimate of potential yield from planted forests in 2005 in Chapter 5 (1.4 billion m^3) is remarkable. It indicates that two-thirds of the world's industrial roundwood production can come from just 7% of the world's forests, or a mere 2% of land cover. Incidentally this supports Clark's (2001) contention of no global wood shortage. With the investment in genetic tree improvement and realizing the potential of biotechnology (Park, 2002; Sedjo, 2004; Sutton *et al.*, 2004; van Frankenhuyzen and Beardmore, 2004; Nehra *et al.*, 2005) – always supposing FSC and other certification bodies don't persist with bans (Strauss *et al.*, 2001) – alongside strategies of concentrating on the best-adapted sites (Fox, 2000) and other aids to improvement, further increases in productivity per hectare are already assured. Further comment is made later about genetic improvement, but Kanowski and Borralho (2004) report 200 tree species subject to one breeding cycle and 60 or so more intensively. Initial improvement yields typically 10% gain where there is small natural variation and rather more where this is higher. Examples of forest-scale gains, as opposed to experimental trials, are still limited, but Evans (2005) attributes a 9% improvement in height growth for fourth-rotation *Pinus patula* to genetic tree improvement. In the UK genetically improved *Picea sitchensis*, the principal planted species, has predicted gains of 8–15% in height and over 20% in diameter (Lee, 1999). Rapid capture of such heritable gains through clonal planting is now widely pursued, for example, in Aracruz, Brazil (Campinhos, 1999) (Fig. 6.1). Of course, tree breeding and biotechnology interventions also aim at delivering improvements in disease resistance, wood properties and other benefits. Increasing yield per hectare from silvicultural intensification is a continuing trend in planted forests for production.

Overall, fibre supply from planted forests is set to increase dramatically and eliminate any lingering spectre of wood shortage globally, if not always locally. Not only will the resource of planted forests largely meet current levels of demand for industrial wood, but, in the medium term, surpluses are possible that can make inroads into and substitute use of non-renewable construction materials that are far more energy-intensive – cement, steel and aluminium (Bowyer, 2004). Such surpluses are highly likely if significant investment occurs in carbon afforestation and reforestation as a climate change mitigation strategy. Not only will much virgin wood be grown but, ultimately, its very best use is a win–win, both to substitute for other materials as a renewable, low energy-consuming alternative and in ways that have long in-use life, as do most construction and furniture uses, and so prolong carbon storage. Indeed, consumer preference is beginning to place a premium not only on competitive prices but also on environmental and social justification in product use. There is every prospect of reversing the trend of the last 50 years, and seeing solid wood and reconstituted wood products regain market share in the construction industry

Fig. 6.1. Harvesting of clonal *Eucalyptus* in Aracruz, Brazil: wide corridors of native forest and protection of riparian zones surround such intensively managed planted stands. (Source: FAO.)

Product quality

But questions must be raised: (i) The roundwood supplied from planted forests is much less varied than has hitherto been extracted from native forest. And not all tree species do well when planted; (ii) Wood quality will change because of fast growth rates, proportionally greater juvenile or core wood, possibly less heartwood, shorter lengths of clear timber and poorer finishing qualities of solid wood because of wider rings from faster growth; (iii) Log dimensions will generally diminish. The large diameters of 'old growth' will give way to smaller sizes, necessitating re-equipping of mills and greater investment into finger joints and similar ways of utilizing small-dimension material; (iv) These three trends will be mitigated by an increase in adoption of re-constituted board and panel products, and by including wood quality parameters in genetic improvement programmes (Lee, 1999; Barbour, 2004); and (v) Fibre supply for pulp will also change, with even less coming from native forest than now (Simula, 2002) and with many changes for the better from advances in technology (Bailey *et al.*, 2004), in particular the uniformity of industrial feedstock that planted forests afford. The industrial resource, based on planted forests, will be different. Offsetting benefits include greater security of supply, uniformity of product – species and sizes, and, obliquely, the benefit of easier certification of planted forests because of demonstrable compliance with regulation standards.

Premium hardwoods

A further question concerns supply of premium hardwoods. In temperate countries long rotations and costly silviculture (Kerr and Evans, 1993; Joyce, 1998) will deter major investment on commercial grounds, though some production will arise from plantings of native hardwoods for purposes such as amenity and biodiversity (Fig. 2.2). In the tropics the picture is bleaker. There are the same issues of long rotation and cost, but also silvicultural problems with the two most important families, namely, shoot borers with most *Meliacaea* (mahogany) (Evans and Turnbull, 2004) and poor establishment and erratic growth of dipterocarps (Weinland, 1998). There is huge potential and opportunities (Salleh Mohd, 2000; Varmola, 2002; Kjaer, 2004), for example with the little-researched genus *Inga* in the neotropics (Pennington, 1997). On the whole, foresters have been content to confine interest and focus research on 'easier' species to grow industrial rather the cabinet grade timber – with, of course, the well-known exception of teak (Pandey, 2000). There appears no immediate change to this outlook.

Outlook

Planted forests are becoming the world's industrial feedstock for wood products. This will require similar inputs as farm crops of high-quality germplasm, site/species matching and, for forestry, a relatively high intensity of silviculture – establishment, protection, management, harvesting and regeneration. Such forests will represent one branch of an emerging dichotomy: that of intensive cropping for industrial end uses in contrast with less intensive management for many non-industrial uses.

6.2.2 The bioenergy future

Burning wood and charcoal for fuel, i.e. using biomass for energy (bioenergy), dates back to the dawn of civilization. As noted in Chapter 2, better recognition of its significance in developing countries emerged in the 1970s, though in the 1980s the 'fuelwood crisis' was overblown. The role of planted forests was not always satisfactory, as many failed woodlots from the 'give 'em eucalypts' era testify. Crash planting programmes of misguided aid packages and loans from development banks were ill-conceived principally by failing to engage with stakeholders (Fig. 6.2). Today these deficiencies are being overcome through effective rural appraisal strategies and participatory approaches, though the need for fuelwood in the tropics and subtropics has not abated (Youngs, 2004). For example, in India, 75% of rural energy comes from biomass (Ramachandra *et al.*, 2004). While in the past native forests and woodland have been the main source, increased reliance on planted forests is likely.

The high dependence on bioenergy in the tropics is similar to the situation in temperate countries 200 years ago before the advent of the fossil fuels, coal, oil and natural gas. But with the diminishing of these fuels and their luxuriant consumption deprecated because of climate change, interest has massively reawakened in bioenergy across the world (FAO, 2008a). And planted forests, including for short-rotation coppice and energy production, will be one of the mitigating options (Berndes *et al.*, 2003; Patzek and Pimentel, 2005). Indeed, Innes (2004) argues that this is a more valuable function – to supply biomass

(a)

(b)

Fig. 6.2. (a) Failed woodlot of *Eucalyptus camaldulensis* in Mali because of poor engagement with stakeholders and termite damage; (b) Successful woodlot of *E. camaldulensis* about 2 km from woodlot (a).

such as wood chips for burning, including co-firing in power generation – than afforestation to sequester carbon (see below). Although concern is expressed about availability of land for biomass, and how the 'poor' are more affected than the 'rich', it is considered the most likely source of renewable energy in the medium term compared with 'high tech' ways of intercepting solar radiation, such as photovoltaic systems (Nonhebel, 2005). Yields range from 5 to 25 dry tonnes/ha/year, depending on species and site. Poplars and willows (Fig. 6.3)

Fig. 6.3. Short-rotation coppice for energy: Willow (*Salix* spp) being used in Sweden, the UK and temperate Europe.

dominate bioenergy silviculture in temperate regions (Dimitriou and Ma, 2005), and acacias and eucalypts in the tropics (Cannell, 2004).

A key issue pointing to massively increased uptake of bioenergy, including bioenergy from planted forests, is their use for gas or liquid fuel production, i.e. in second generation (2G) biofuels (WBCSD, 2007; Evans, M. C., 2008, personal communication). Ligno-cellulosic (wood) pyrolysis or distillation, the latter for methyl and, particularly, ethyl alcohol (Zerbe, 2004) are becoming attractive as sustainable energy alternatives. Development of cost-lowering technologies for the processes could bring these fuels on-stream at economically competitive rates from planted forests and without some of the downsides – displacement of food production, massive clearances of native forests – associated with first generation (1G) biofuels derived from cereal grain, sugar or vegetable oils. However, efficient production of 2G biofuels from wood may be years or possibly decades away (Evans, M. C., personal communication).

Accounts are available outlining bioenergy potential and use for many countries; those in the list below are illustrative.

- Australia (Fung *et al.*, 2002)
- Belgium (Garcia-Quijano *et al.*, 2005)
- Canada (Hall and Richardson, 2001)
- China (Junfeng and Runqing, 2003)
- Czech Republic (Lewandowski *et al.*, 2006)
- Denmark (Jorgensen *et al.*, 2005)

- Germany (Hoffmann and Weih, 2005)
- India (Sudha *et al.*, 2003; Hooda and Rawat, 2006)
- Malaysia (Koh and Hoi, 2003)
- Philippines (Elauria *et al.*, 2003)
- Poland (Ignaciuk and Dellink, 2006)
- Sri Lanka (Perera *et al.*, 2003)
- Sweden (Hoffmann and Weih, 2005)
- Thailand (Sajjakulnukit and Verapong, 2003)
- USA (Cook, 2000)
- Regional accounts – boreal conditions (Weih, 2004), temperate conditions (Dickmann, 2006).

While new planting is expected, increased use of residues in the forest and at the mill will also occur as the costs and benefits justify. As with all woody biomass, traditional features of timber quality are less important and this demand can be turned to advantage to carry out improvement cleanings and thinnings – tending operations – in planted forests and find a market for otherwise unwanted material. One such innovative possibility could be 'biochar', the locking up of organic carbon in forms that do not readily break down, through low-temperature pyrolsis, and used to augment soil carbon stocks (Lehman, 2007). Planted forests will be part of the bioenergy future.

6.2.2 Will forests be planted to store carbon?

Trees accumulate carbon during their life and can often enrich soil surface and subsoil carbon stocks with recycling of organic matter, their root systems and formation of peat. Thus appropriate afforestation and reforestation are seen as weapons to slow the rise in atmospheric carbon dioxide and in turn help mitigate global warming (Fig. 6.4). However, exaggerated claims have been made, and it has even been argued that because planted forests mostly grow faster than native forests, it is best to clear them and plant trees. Such action will rarely be carbon positive, nor ecologically sound or socially justified. Indeed, the best single contribution the forest sector can make to mitigate climate change is to prevent the land-use change that is deforestation (Kirschbaum, 2003; Innes, 2004; Alvarado and Wertz-Kanounnikoff, 2007). Even forest degradation that lowers carbon stocks is increasingly seen as significant, as is the concern that tropical forests may not be fully sequestering the carbon that climate change models predict (Fox, 2007). Reducing carbon emissions from both deforestation and degradation, the so-called REDD initiative, are now seen as crucial to any climate change mitigation strategy (Dutschke and Wolf, 2007).

Comparison of carbon stocks – biomass and the soil's carbon pool – generally shows the following relativities: native forest>planted forest>arable crops (Shan *et al.*, 2001; Garten, 2002; Guo and Gifford, 2002; Lasco, 2002; Smith *et al.*, 2002; Chen *et al.*, 2005; Cerli *et al.*, 2006). The point of 'arable crops' is that soil carbon stocks under some traditional pastures have been found to be similar to or exceed that of planted forests (Guo and Gifford, 2002; Mendham *et al.*, 2003; Maraseni *et al.*, 2007), though not necessarily poor grassland

Fig. 6.4. High standing volume in excess of 1000 m^3/ha of *Eucalyptus grandis* in KwaZulu Natal, South Africa. About 500 tonnes per ha of carbon are stored in the trees and their roots, but the net capture will depend on comparison with the carbon stocks of the site before it was planted 28 years ago.

(Heriyanto, 2003). Modelling these kinds of relationships allows creation of a carbon flux history (Houghton, 2003) – essentially the interplay of deforestation, agricultural land use, and afforestation and reforestation. For China, Houghton and Hackler (2003) reconstruct land-use changes and show peak emissions from rapid deforestation in the 1950s to net sequestration since the 1990s because of diminished forest clearance and expanding afforestation, i.e. from source to sink. For sub-Saharan Africa, Houghton and Hackler (2006) examined changes in five land types and calculated that the subcontinent has been an increasing carbon source since 1900. By the 1990s it was contributing 15% of the global net flux of carbon from land-use change. The clear message is: don't deforest or degrade forests, but seek to increase forest cover.

One issue of concern where afforestation takes place on carbon-rich soils, such as peats and peaty-gleys common in temperate and boreal regions, is whether establishment operations such as ploughing accelerate breakdown (oxidation) of organic matter and liberate more CO_2 than the planted trees would ever store. The situation is not clear-cut. Some models of the forest carbon sink assume a steady oxidizing of organic matter at 3% (Cannell and Dewar, 1993), while Hargreaves *et al.* (2003) reported that recently drained peat switched from being a carbon source to a sink after 4–8 years, once colonization and canopy close had occurred. Byrne and Farrell (2005) concluded that on blanket peats losses of soil carbon are largely compensated by tree carbon uptake. Establishment processes are a concern, but may be of lesser importance compared with other threats to organic rich soils, especially warming of the tundra. Nevertheless, optimizing water management in peatlands, i.e. reducing or eliminating drainage, is according to Parish *et al.* (2007) the single highest priority to combat carbon emissions. Thus afforesting blanket peats may be best avoided.

Inclusion of tree planting in the Kyoto Protocol (Clean Development Mechanism – CDM) acknowledges the role deliberate afforestation can play. The first scheme approved under the mechanism was in China. Estimates of the impact of such carbon planting at country level have been made. For example, Chile's planted forests of *Pinus radiata* were found to compare favourably with other forest types as a carbon sink (Espinosa *et al.*, 2005). Ireland's currently rapid afforestation accounts for 22% of the country's emission-reduction commitment under the Kyoto Protocol (Byrne and Milne, 2006). The UK's planted forests are a carbon sink sequestering about 1.5% of annual emissions (Cannell and Dewar, 1993). And in the USA, Wong and Alvalapati (2003) predict a small but positive benefit. The question remains, however, will afforestation and reforestation for carbon sinks come about on a large scale? Trading in carbon has been slow to develop and while opening up of carbon markets will alter planted forest management (e.g. Stainback and Alavalapati, 2005), designing markets to benefit all, including those in developing countries, is still in its infancy or speculative (Niesten *et al.*, 2002; Olschewski and Benitez, 2005). Although valuations for tonnes of carbon sequestered vary, they do not need to be exorbitant to exceed income from conventional tree harvesting in planted forests (Greig, 2007). It seems inevitable that the increasing emphasis placed on the environmental benefits of forestry, including of planted forests, will create a market for and payment of such services (Neeff *et al.*, 2007). Devising effective policy strategies will be complex (as in the example of US work; Richards *et al.*, 2006).

Of course, using afforestation in these ways could bring conflicts of other kinds, particularly social (Smith and Scherr, 2003) and over water use (Farley *et al.*, 2005). But, if as seems probable, planted forests do become widely established for their carbon, they will also become a source of wood products and offer a win–win opportunity, especially when products such as lumber store carbon for decades.

6.3 Some Social Roles of Planted Forests

6.3.1 Poverty reduction and sustainable livelihoods

The Millennium Development Goal of eradicating poverty naturally raises the question: what role does planted forest play? Being overwhelmingly a rural development – while recognizing the importance or urban and peri-urban plantings – and because many poor people, including subsistence farmers and many landless and refugees, are rural dwellers, can planted forests alleviate this privation? Conversely, and to put it bluntly, can planted forests also exacerbate poverty, for example by displacing people from their land or disrupting local socio-economic patterns? The issue also bears on the urban poor, since many people who live in towns and cities rely on wood for fuel and building materials. Poverty reduction and sustainable livelihoods are huge subjects and very important. This chapter can only briefly examine issues relating directly and indirectly to the impacts of planted forests: it draws particularly on Garforth and Mayers' (2005) book *Plantations, Privatisation, Poverty and Power*. For convenience, issues are loosely grouped according to scale.

National and commercial development of planted forests

Development of planted forests will necessarily impact a nation's or region's GDP and their contribution is often underestimated (Garforth *et al.*, 2005b). The relatively intensive management of such forests leads to inward investment, including for infrastructure and often social services, with consequences for local and regional economies (Evans and Turnbull, 2004). Direct employment and other benefits will plainly accrue, but issues are never simple. For example, there may be migration of labour to projects, or questions of land ownership and traditional rights, and frequently there is an interplay between different land uses (and drivers of land-use change), including pressures to deforest and pressures to plant trees. This is recognized by the global partnerships on forest landscape restoration that aim to enhance the contribution of forests and trees to benefit people's livelihoods, local communities and ecological integrity, including biodiversity. Planted forest development cannot be viewed in isolation and nor can its impact on poverty; it is but one player across several sectors and disciplines.

While large-scale deforestation can cause massive disruption to local people, development of planted forests can also interfere with traditional land-use rights or people find themselves becoming squatters (Evans and Turnbull, 2004). Attempts to address this have sought to include local people in ownership and management, such as in the Fiji Pine Scheme, India's Joint Forest Management initiatives (Rishi, 2002; Saigal, 2005), or Indonesia's community–company partnerships (Nawir and Santoso, 2005). It is critical to engage in dialogue with communities in planning and practices that impact upon them.

A complementary strategy is to recruit smallholders into tree planting and build part of an industrial resource that way – outgrower schemes. One of the conclusions from the FAO Global Planted Forests Thematic Study is a clear trend in this direction. As long ago as 1996, a global survey showed that more than 60% of pulp-producing companies sourced some of their product from outgrowers (Garforth *et al.*, 2005a). For example, in South Africa and in the Philippines such smallholder forestry, where planted woodlots are grown under contract to supply mills, has created a significant source of timber and employs tens of thousands of people (Dlomo and Pitcher, 2005; Bertomeu *et al.*, 2006). Evidence from such schemes suggests that relatively equitable benefits can flow to poor people but, for South Africa, each household needs at least 6 ha to be viable (Garforth *et al.*, 2005a). As a model to replicate elsewhere and as a tool to alleviate poverty, it is attractive but plainly confined to those with access to some finance and land, and an enabling political and legal environment, i.e. not available to the poorest of all.

Also at issue is whether such smallholders have access to top-quality planting stock, market information, technical support and other enabling policies, allowing them to compete with commercial operations (Carle, 2007). Some international agencies are specifically targeting this concern (e.g. The Gatsby

Charitable Foundation) by ensuring the supply of the best possible genetic stock. There is also the question of knowledge and skill – or lack of them – whether of silviculture and leading, for example, to a preference to work with native (and familiar) species (Piotto *et al.*, 2003) or of market intelligence – access to and knowledge about the market (e.g. in China's rapidly evolving economy and rural to urban demography; Weyerhaeuser *et al.*, 2006).

Alongside developmental issues is the pervasive impact of global climate change and the expectation that poor people will suffer the most. Not only are they least able to adapt their lifestyles, but they are usually in the most vulnerable locations, threatened by rising sea levels or worsening droughts. Climate change will impact protection forests or efforts to combat desertification through planted forests and so increase risks to local people (Innes and Hickey, 2006). And, if carbon sequestration by forests, including planted forests, becomes one of the strategies of the international response to climate change, will local people be able to benefit and what are the trade-offs (Smith and Scherr, 2003)?

Large-scale investment in planted forests can undoubtedly benefit many poor people, and with a favourable and creative policy environment can contribute to alleviating poverty. As Turnbull concludes concerning socio-economic impacts of China's extensive eucalypt afforestation programme, 'although plantation development has contributed significantly to poverty alleviation, it is probable that greater benefits accrue to higher-income groups' (2007, p. 67). It is clear that investors need to recognize and encourage activities with local communities wherever practical to do so.

Medium scale: direct provision of villagers' needs

Traditional planted forests for timber production usually only meet the needs of poor people in limited ways beyond that of employment and informal enjoyment of by-products (Garforth *et al.*, 2005a). Hence investment in planted forests specifically to provide fuel, building, fencing and other materials for domestic consumption has been a crucial tool in aid – they are 'a common invention in developing countries' (Kohlin and Amacher, 2005). As well as yielding the desired commodity, they can bring welfare improvements of time saving for households and villagers. In Bangladesh such social forestry is described as extremely successful, compared with traditional forest management, increasing forest cover by more than 40,000 ha and strip plantings even more, generating income for participants and benefiting thousands of poor people (Muhammed *et al.*, 2005). In Nepal policy shifts in support of local people have been 'tremendous', but the actual pace of community forestry development slow (Acharya, 2002). In Ethiopia numerous initiatives, many achieved through food for work programmes, have created landscapes of woodlots and small-scale plantings (Evans and Turnbull, 2004), but success was in part conditional on a form of collectivization in the 1970s and 1980s.

There is an abundance of literature on rural development forestry and, as pointed out, community and social forestry have by no means always been successful. Medium-scale developments of planted forests have created many new resources, but are not a panacea for the ills of the poor.

Fig. 6.5. Women carrying firewood and building poles from a planted woodlot in Lesotho.

Family and individual needs

Provision for domestic livelihood needs of food, shelter, shade and fuel (as well as education, health and security) (Fig. 6.5) need not be confined to initiatives at the village or community level. Mention must be made that tree planting near homesteads and in compounds, along field boundaries, beside tracks and in other ways – trees outside forest – can directly help. The countless home gardens of the moist tropics (Belcher *et al.*, 2005), the carefully guarded pollards of the semi-arid subtropics (Evans and Turnbull, 2004), the adoption of agroforestry practices, and the widespread gathering of non-timber forest products (Cavendish, 2001) all indicate a key role played by the individual.

It is not surprising that a key role played by numerous forest nurseries – the traditional 'extension' nurseries – is to supply a wide variety of trees for growing food, for wood products, for enriching soil through nitrogen fixation, for shade, for fuel: the list is nearly endless. It is one more small input to add rather than take away resources from local people who, though poor in cash terms, can invest their time in cultivation. Their labour creates some capital – the growing trees – and in many cultures is as good as money in the bank.

The key to achieving poverty alleviation

The experience of the last 30 years has revealed that, above anything else, for planted trees and forests to help there must be full and sufficient engagement with, and participation of, all stakeholders, but crucially the 'beneficiaries' themselves (Mayers and Bass, 2004). Foisting solutions on people rarely works.

- Good governance must be in place that permits all stakeholders to participate in decision making, encourages collaboration and shared ownership (Purnomo and Guizol, 2003).

- Land and tree tenure must be clear and transparent regarding who owns what.
- Local people must select and implement interventions (Saxena *et al.*, 2002).
- Poor people must derive income from their participation.
- Sufficient rights must be vested in the beneficiaries.
- Free and informed consent must be given.
- Best practice guidelines must be available and communicated.

These bullet points demonstrate that, ultimately, it is empowerment of poor people that is the critical step leading to opportunities and openings. And planted forests have a role to play.

6.3.2 Amenity, recreational and related services

Many planted trees and forests are enjoyed for what they are: objects of beauty, places to walk, somewhere to be quiet. Whether these emotional, spiritual and aesthetic values are greater for native forests or species is not often asked. While there is little debate over which is preferable for biodiversity and wildlife conservation, many planted forests are enjoyed for amenity and recreation. Indeed, the great arboreta of the world, such as Britain's Kew Gardens or National Arboretum at Westonbirt, attract visitors by the thousand and are necessarily full of exotic species.

The key issues appear to be age and diversity as surrogates for naturalness rather than whether the trees are a native species or not. London's famous and much loved plane trees (*Platanus* × *acerifolia*), New York's Gingkos (*Gingko biloba*), or Tshwane's (Pretoria's) Jacarandas (*Jacaranda*) that line the streets and adorn the parks and squares, are all exotic. Examples can be multiplied city by city – Nairobi, Tehran, Singapore to name three that are well-known for their trees. Much urban woodland, whether of native species like Brussels' wonderful beeches (*Fagus sylvatica*) in the Forêt de Soigne, or of exotic species like the fabulous and much visited redwood (*Sequoia sempervirens*) grove in Rotorua (New Zealand), can be equally popular and valued. Urban woodland is usually regenerated by planting to ensure rapid establishment to maintain the treescape (Nilsson *et al.*, 2001a). Planting as such is rarely perceived as an impediment.

But the question of age and diversity is significant. Millions visit the UK's planted forests; indeed, their value for non-market services such as outdoor recreation far exceeds that from production of wood products, but what visitors dislike is uniformity – monocultures of species and of ages. Deliberate interventions to extend rotations and grow large old trees – for recreation and wildlife (Humphrey, 2005), early fellings to add structure to an otherwise uniform forest, and supplementary planting of native species are all pursued to enhance landscape and amenity values (Gill and McIntosh, 2001).

Planted trees and forests are well able to serve these roles, provided sympathetic management offers variety of size, age, and of internal and external landscape. Appropriate interventions throughout the life of the stands, from good design of new forests and their layout to timely fellings, all add to enjoyment. This is especially so in peri-urban and similar much-frequented woodland and forest.

6.3.3 Can planted forests slow deforestation?

This question is more of a synthesis of roles already outlined, as well as those associated with protective forest below, and is addressed specifically because it is often asked. As a question, it is shaped in terms of a role planted forests could play for the benefit of society, hence including it now, rather than a more specific purpose. The issue tends to focus on the tropics, but in many countries where forest cover has doubled or tripled in the last 100 years through afforestation (Denmark, Ireland, the UK), or where planted forests overwhelmingly provide the industrial feedstock for wood products (Chile, New Zealand, South Africa and even the USA), it has become possible to divert pressure away from, and so conserve remaining areas of, native or semi-natural woodland and forest. Does this have a wider relevance?

As this book describes, timber production is fast shifting to a planted forest base, but logging for timber has not been the exclusive or even dominant driver of tropical deforestation. Agricultural development, transmigration of people, mining, even wars have all at times been far more important (Ghazoul and Evans, 2004). So although transferring wood production to planted forests is occurring, it is unlikely to be critical beyond the admittedly hugely important matter of public perception. For many people it will appear simple: if planted forests are meeting the world's demand for wood, there's now no need to deforest the tropics. In this sense, the expansion of planted forests could have a massive impact.

Specific roles of planted trees and forests best address pressures to deforest arising from displaced people or those reliant on subsistence farming. The two principal interventions are creation of buffer zones of planted forests and promotion of agroforestry practices at the forest margin. Both create a transition environment in terms of microclimate and ecology, as well as economically, which can deflect pressures to use native forest. Further comment is made later.

The converse must also be asked: do planted forests sometimes cause deforestation? While today clearing native forest simply to provide sites for planted forests is deprecated, Chapter 2 showed that one of the reasons for tree planting in the past was to replace slow-growing native species with faster-growing, often exotic, ones. The history of forest conversion in Germany 150 years ago, and in the UK 50 years ago must conclude that planting caused some loss of native forest – though currently this process is being reversed through restoration strategies (see below). Recently in New Zealand expansion of planted forests was singled out as the most important driver of deforestation (Ewers *et al.*, 2006). While attention of forest policies focuses on planted forests, there can be neglect of native forests, even by default, if only in the relative amounts of resources allocated.

Few would disagree with Mather's conclusion that 'planting trees would in itself not save the forests, but it is a good beginning' and it is 'certainly difficult to see the global forest being saved if trees are *not* planted. It does not itself offer a solution or a "technical fix" to the problem of shrinking forest resources' (1993, p. 207). Establishing planted forests is not going to be a primary tool to slow deforestation, just one more help. But, importantly, a help too in changing the public perception in the West that using wood destroys rainforest when in fact most paper and wood products come from sustainably managed planted forest. Greater awareness of this could be crucial, and sharpen the focus on the real drivers of deforestation.

6.4 Environmental Services

6.4.1 Can planted forests help mitigate climate change?

The role of planted forests for bioenergy and carbon sequestration were under 'Production' because they are also likely to supply forest products: clearly though it is the huge threat climate change poses that promotes their development. If planted forests are fast satisfying world demand for timber and if new planted forests for bioenergy and carbon are established, their greatest contribution could be to provide feedstock for renewable fuels and for alternative construction materials to substitute energy-intensive aluminium, concrete and steel and do so in a use which often stores carbon for decades. To the extent that these developments occur, planted forests are a 'win–win'.

And, as just outlined, planted forests may also play a role in helping to alleviate deforestation, which is one of the critical sources of carbon dioxide, and prevent this damaging land use change. But establishing planted forests is itself a land use change and it is unclear whether in terms of global warming directly, as opposed to carbon storage, it is always necessarily positive. The albedo of forest is low and so absorbs more of the sun's radiant energy than some other land types. For example, in high latitudes more is reflected from open snow fields than forest under snow (Innes, 2004).

Planted trees and forests also have a more local role in mitigating the consequences of climate change. A familiar example is the establishment of shelterbelts and windbreaks and the ameliorative effect they have (Gardner, 2004). The same can be said for trees and woodland in towns and cities which, while adding 'green space' and all its therapeutic benefits, ameliorate the local climate (Nilsson *et al.*, 2001b).

As noted earlier, there is also the public's perception that planting a tree is a good thing. In many countries it is the commonest response by people wanting to counter global warming or clearance of rainforest. Though possibly ill-informed and perhaps a distraction from more important issues, it is a powerful motive. From heads of state downwards, tree planting is the great act of commemoration and of doing good. The enduring appeal of Arbor Day in the USA, which dates back to 1872, is evidence enough and even then the motivation was that of helping to change the climate (Williams, 1993)! Channelling interest in tree planting to raise awareness of greater issues to do with climate change, is a media role that planting and planted forests may increasingly come to play.

6.4.2 Protective afforestation

The role trees and forests play in protecting the environment cannot be underestimated – hence the broad classification of forests into productive and protective in FRA 2005. Forests are a key element in the landscape. They provide a natural buffer that assists in maintaining ecological balance and supplying raw materials to communities as well as providing protective services. Forest practices need to ensure the health and vitality of forests if they are to be managed sustainably and fulfil their protective role, especially the protection of water resources, soil protection, buffering local climate through control of wind and other conservation and recreational functions (Göttle and Sene, 1997).

Tree cover usually reduces soil erosion, slows wind speed, traps airborne sand and dust particles, moderates the force of rain and slows water runoff after heavy rain. Also, planting trees on degraded land, mining sites or sand dunes can be the first significant step towards soil rehabilitation, phytoremediation and land reclamation and, for example, assist in combating desertification. In protective afforestation, or any forest reserved and managed for protection, timber production is of secondary importance, a fact which markedly affects silviculture and management. The protective role of the trees, in particular the surface litter they produce and accompanying vegetation become the dominant consideration in all decisions, such as what species to plant, whether to thin, how to regenerate the forest, whether to allow firewood collection and livestock grazing and so on. However, many of the sites most urgently needing protective tree planting are often those which are inhospitable and where establishment of planted forests is difficult. And it must be stressed that tree planting alone is rarely a sufficient protective measure. In the case of soil erosion much myth and a lot of hope centre on what the planting supposedly will achieve when, in reality, it is the associated removal of grazing or the construction of terraces or the control of land use, such as exclusion of fire or prevention of litter raking, which accompanies the tree planting that is crucial (Thorne, 1989). In the conservation and protection of mountain watersheds, integrated planning and management is required to maintain optimum interrelations between forests (including planted forests) and water, wildlife and soils, as well as a high level of participation of local communities and other stakeholders (Fernandez, 1997).

It used to be said that planted forests have been little used in protection forestry (Evans and Turnbull, 2004). They were limited mostly to shelterbelts on farmland and around towns or to stabilize sand dunes. Maintenance of native forest cover in upper watersheds and mountainous regions was the priority for protection and this, of course, must continue. What is now clear is that many such protection forests are kept well-stocked by replanting and fit into the semi-natural planted category of FRA 2005. In addition, deforestation or degradation of many native forests has already occurred so tree planting and/or management of secondary forests for protection are the only options (Jong *et al.*, 2001). Planted forests are usually the only possibility on bare or degraded lands. They have also been used by some countries to compensate for loss of timber supply when logging bans were introduced in natural forests to protect watersheds (Durst *et al.*, 2001).

Examples of protective afforestation programmes

Large afforestation programmes to combat deforestation, on deforested hills and on coastal sand dunes of China, have an important soil protection role as well as for production of timber. Establishment of shelterbelts began in various parts of China in the 1950s (Song, 1991). Sand dunes along the coast of the South China Sea were stabilized with *Casuarina equisetifolia* plantings, resulting in environmental and livelihood benefits (Turnbull, 1983). In 1989 the Chinese government sponsored the Yangtze Shelterbelt Programme to conserve forests in the upper and middle reaches of the Yangtze River and in 1998 the Natural Forest Conservation Programme's logging ban in native forests was accompanied by a plan to establish 21 million ha of planted forests for timber in the upper reaches of the Yangtze River and the upper and middle reaches of the Yellow River (Yang, 2001). Of comparable scale is the Three North Shelterbelt Programme using mainly hybrid *Populus* spp. (Carle and Ma, 2005).

The Government of India's Forest Conservation Act of 1980 was designed to reduce the annual loss of forest area and at the same time a Social Forestry Programme was initiated to reclaim degraded forests and village commons and to meet fuelwood demands. An estimated 100 million ha of degraded land (forest, village commons and marginal farmland) has little vegetative cover and is subject to soil erosion (Ministry of Environment, 1990). The mean area afforested annually from 1980 to 1998 was 1.4 million ha, making it one of the largest afforestation programmes in the world (Ravindranath and Hall, 1994). Although the benefits of this programme to landless rural population are debatable, it undoubtedly increased forest cover and assisted soil conservation. Protection from fire, grazing and fuelwood collection from planted forests resulting from the Social Forestry and Joint Forest Management Programmes has encouraged natural regeneration of local species in these plantings, providing an opportunity to develop secondary forests with greater conservation value (Bhat *et al.*, 2001).

Major tree-planting projects as part of soil protection measures have been made in many parts of Africa (e.g. Ethiopia) and Asia (e.g. Philippines). Also countries bordering the Sahara to the north and east, most notably Algeria, Libya, Morocco, Sudan and Tunisia, have established planted forests to combat desert encroachment (see section 6.4.4).

6.4.3 Do planted forests adversely affect hydrology?

When water from a catchment is used for drinking, irrigation or generation of hydroelectric power, management of the watershed is critical, since it can affect both the quality and quantity of water supplied. The inorganic and organic constituents of the water reflect the mineralogy of the watershed, the character of the precipitation and the nature of the vegetative cover (Hewlett, 1982). Changes in vegetative cover by deforestation or tree planting therefore have significant effects on the hydrology of a watershed. What these effects are and their value can be contentious. Calder suggests that reports that forests increase runoff, regulate flows, reduce erosion, reduce floods and improve water quality are 'seen to be either exaggerated or untenable' (Calder, 2002, p.38) when examined critically. He has pointed out that foresters and hydrologists often have very different perceptions regarding the hydrological role of forests, with hydrologists concerned that trees intercept more rain during wet periods, and because of deeper root systems deplete groundwater by transpiring more water in dry periods (Calder, 1996). He also suggests that it is unrealistic to attempt to generalize forest impacts as they affect extreme flows. Also, some negative impacts related to floods and erosion may be more associated with forest management operations such as logging, site preparation and roading, rather than the presence or absence of the forests themselves. Non-hydrologists may, indeed, have over-simplified the impacts of forests in water catchments, especially when attempts are made to generalize from reports of events and research that may be very site-specific. But because water availability is one of the great issues of the present century and will be exacerbated by climate change, the relationship between forests, and especially planted forests, and water supply needs appraising.

Reviews of water use by forests, including planted forests, will be found in Bruijnzeel (2004), FAO (2008b), Scott *et al.* (2004) and Vertessy *et al.* (2003).

Water quality

Tree and ground cover can greatly reduce soil erosion. In watershed management this not only prevents loss of fertile topsoil, but means that water draining from forested or well-grassed catchments will be largely free of sediment. And, as well as little loss of soil from the eroding land itself, little or none will be transported and deposited elsewhere. Though Finlayson (1998) has highlighted the difficulty of establishing a useful relationship between erosion and sedimentation, sediment deposition in reservoirs and irrigation channels is a widespread and costly consequence of deforestation in many tropical countries.

A second effect of forest cover in catchment areas is that alternative, often more polluting, land uses are excluded. Water draining from a largely forested catchment is usually cleaner than from one where land is used for grazing or arable farming, which may include the pollution hazards of fertilizers (especially nitrates and phosphates), pesticides, and human and animal wastes. For example, in Malaysia the quality of stream water was increasingly degraded as it passed from undisturbed forest, through a swamp area, logged-over forest and agricultural land (Yusoff *et al.*, 2001). However, these benefits of forest cover cannot necessarily be claimed for managed planted forests established on watersheds.

Many operations associated with planted forests cause disturbance and exposure of soil through ground preparation, planting, control burning, forest grazing, thinning, timber harvesting and road construction. Indeed, this is one of the reasons why, in forests planted for a protective function, timber production is of secondary importance to that of leaving tree and ground cover intact. It is also why chemicals such as herbicides and insecticides are avoided. The quality of water draining from a forested catchment is very much under the forester's control and increasingly codes of best practice are being developed to ensure environmental protection, e.g. the UK's Forest and Water Guidelines. Awareness of the physical, chemical and biological indicators of water quality from catchments, such as turbidity, pH level, aquatic organisms etc., is crucial (Walker and Reuter,

1996). In general, however, most of the undesirable effects of forest management on water quality can be much reduced by leaving buffer zones of undisturbed vegetation next to watercourses, carefully planning location and time of operations, and through holistic management keeping the protective function uppermost.

Influence on water quantity – afforestation and streamflow

Forest cover, compared with open land, affects both the pattern and total quantity of water discharge from a catchment area. Both have important implications for flood control and land-use practices downstream. Trees planted for commercial or protective purposes may use more water than the crops or grass they replace. The impacts will vary on different sites and are influenced by the nature of water flows, landscape features, area and density of plantings and management (O'Loughlin and Nambiar, 2001).

Pattern of water discharge

In general, loss of vegetation leads to more severe flooding, but it does not follow that planting trees will automatically eliminate flooding problems. A forested watershed generally has lower peak flow rates after a storm than one recently denuded of forest, under arable farming or overgrazed, so forests exert some regulating influence on the flow of water. The mechanism is the same as that which reduces soil erosion: forest cover slows the movement of water through part of the hydrological cycle. This slowing principally arises from the combined effect of understorey vegetation and a well-developed litter layer (Scott *et al.*, 2004), which is why many *Eucalyptus* spp. and teak (*Tectona grandis*) are unsuitable for such planting. The result is better infiltration of water into the soil, reduced surface runoff, and therefore slower drainage from the catchment into streams and rivers and hence reduced velocity of flows.

While there is observational evidence that stream flow from forested land is prolonged during dry periods, e.g. in the Ngoronit stream catchment in northern Kenya, and it is a common observation in the tropics that streams that once flowed from forested catchments often dry up following deforestation, there is increasing evidence that planting fast-growing trees on grasslands diminishes streamflow, especially in the dry season low flows (Bruijnzeel, 1997; Calder, 2002).

Studies that show planted forests reduce runoff and peak flow rates, and so prevent flooding, mostly apply to 'normal' heavy rainfall events and not a deluge when huge amounts fall in a short period. Planting trees in a catchment will rarely guarantee that there will be no floods. Trees have a role to play, not least in reducing soil erosion and landslides, but their effectiveness in flood control should not be over-emphasized.

Total water yield

It is common practice to stand under a tree to keep dry (interception) and to plant trees, especially species of *Eucalyptus*, to lower water tables (interception and transpiration). Moreover, when forest is cut, it is frequently observed that the water table rises and the soil surface becomes wetter for a time, since interception and transpiration are both reduced. Thus, with trees affecting hydrology, the consequence of afforestation in a watershed on water yield must be considered, especially in seasonally dry areas and where soils are shallow and store little water.

Trees, like nearly all plants, lose water through transpiration. Because trees root more deeply than most grasses and herbs, they are able to draw on reserves of moisture and continue transpiring at a high rate for longer during a dry period. This effect, combined with evaporation of water intercepted by the canopy, usually results in an afforested watershed consuming more water than one that is not, and therefore in less water draining from it. The reduction can be crucial in arid areas where all water is at a premium. However, it is dangerous to over-generalize.

Eucalypts have a reputation for high levels of water use, but critical evaluation indicates very large differences between species in both stomatal responses and rooting patterns. When water becomes limiting, most eucalypt species respond by closing their stomata and so transpire about as much as other tree species, but a few species have less stomatal control and greater transpiration rates (Calder, 1992). Some species, such as *Eucalyptus camaldulensis*, even show provenance variation in their stomatal response to water stress, with the well-known Petford provenance responding to stress by stomatal closure but not shedding leaves, whereas the Lake Katherine provenance responds by leaf shedding (Gibson and Bachelard, 1994; Gibson *et al.*, 1995). This suggests that *E. camaldulensis* from the dry tropics depends more on reduction of leaf area than control of transpiration to conserve water. Where eucalypts have access to the ground water table, transpiration rates may be very high indeed, especially in hot, dry conditions (Calder, 1992). This explains why some species can rapidly lower groundwater levels and dry out the ground.

However, measurements of water use by vegetation can only be properly interpreted with reference to the location and period in which the measurements are made. In studies in Australia, Pakistan and Thailand, estimated annual water use by planted *Eucalyptus*, *Acacia*, *Casuarina* and *Prosopis pallida* ranged from 300 mm to 2100 mm. On a given site, water use differences between species largely reflected growth rates, in terms of sapwood area per hectare (Morris, 1997; Mahmood *et al.*, 2000). Availability of water and the evaporative demand greatly influence transpiration and may override differences in tree species, size of tree, water quality and other factors. For example, in the UK the data of Calder *et al.* (1997) from planted forests of beech (*Fagus sylvatica*) raised concerns about impacts on water resources of planting broadleaved woodland in lowland England. However, this conclusion has been challenged because the data neglected to account for continued moisture recharge, i.e. availability of water, from the underlying chalk geology (Nisbet, 2005; Roberts and Rosier, 2007). The water balance of agroforestry systems is even more complex than for tree monocultures and currently little is known about the way water is partitioned in them (Wallace, 1996).

Many whole catchment studies have been carried out. In general, as the forest canopy gradually increases from the time a planted forest is established, annual evapotranspiration increases (i.e. with stand age). For example, in a high rainfall area of New Zealand, a native mixed evergreen forest was converted to planted pine forest in the mid 1970s. For the first 4 years, water yields from catchments were 49–74% higher, depending on harvesting method. Flood peaks also increased significantly after harvesting, but within 15 years annual water yields had returned to pre-harvesting levels (Fahey and Jackson, 1995). A similar effect was observed in French Guyana, where runoff increased 60% after primary rainforest was cleared and eucalypts planted. However, after about 5 years the runoff was similar to pre-eucalypt establishment and after 6 years it was 10% less (Cossalter *et al.*, 2003). The situation may be different in catchments where grassland or sparse vegetation is planted with pines or eucalypts. Results of long-term experiments in South Africa showing reductions in annual water yield from grassland and scrubland sites planted with *Pinus patula*, *P. radiata* and *Eucalyptus grandis* and are summarized in Scott *et al.* (2004).

In South Africa it is recommended that where the water is needed and where alternative sources are not available, sites outside the humid forestry zones, or with long dry seasons, are not planted (Fig. 6.6). It is also their practice not to plant trees close to streams and rivers to avoid copious water use by trees and to encourage development of vegetation with relatively low transpiration rates, especially in dry winter months (Dye, 1996). Legislation known as the 'Afforestation Permit System' now regulates planted forests and tree planting development in South Africa, primarily to protect natural water resources (van der Zel, 1995).

Like South Africa, Australia is a dry country with an active planted forest programme. Experience of the impact of planted forests on river flows in Australia has been summarized by Vertessy *et al.* (2003). They note that the fraction of the catchment area planted, where the planted forest is positioned within the catchment, and variations in stand age and site productivity all impact on water yields. They suggest careful planning can minimize negative impacts and that the beneficial environmental effects of planted forests should be carefully considered

Fig. 6.6. In South Africa, competition for water resources between agriculture and forestry has led to planting restrictions near streams and rivers and water-use tariffs on some commercial timber-growing.

in the formulation of policy and legislation directed at management of their development.

In a global analysis of the effects of afforestation on water yield, Farley *et al.* (2005) suggest that water shortages will often be intensified by tree planting and there may be critical trade-offs in some locations between different environmental impacts, e.g. large-scale planted forests for carbon sequestration. Reduction in water yields from afforested catchments must be kept in perspective as in many parts of the world the benefits of careful tree planting to assist erosion control and flood prevention outweigh the disadvantage of some reduction in water yield. Clearly, when establishing planted forests in areas where water availability is critical, the planning of land uses should consider overall ecosystem processes and socioeconomic issues (O'Loughlin and Nambiar, 2001).

A summary of impacts

1. To minimize erosion and sediment release, while it is best to retain undisturbed native forest and vegetation, including grassland, where tree planting is carried out it is essential to minimize soil surface disturbance, and to encourage understorey vegetation and a well-developed surface litter layer.
2. Planted forest for protection must be integrated with other land uses and managed sustainably. Planting trees is only one component of better land-use practices in catchments that complements control of cultivation and grazing, engineering structures, such as check dams and terraces, and buffer zones to moderate part of the water cycle.
3. Afforestation of land where planted trees replace grassland, scrub or degraded land will reduce water yield, peak and base flow rates, though where there is good infiltration base flows may be prolonged. Small to medium-sized stormflow events and consequent flooding may be reduced.

6.4.4 Desertification

Decline in vegetative cover aggravates any tendency to desertification in an arid or semi-arid region. Logically, therefore, one of the goals in reclaiming land in danger of becoming desert is to replace lost trees and forest, and, in such a harsh environment, direct afforestation or introduction of agroforestry practices (Baumer, 1990) is usually the only way. In 1977, the United Nations conference on desertification advocated tree planting programmes to establish shelterbelts, fuelwood plantings and woodlots, and to stabilize dunes. The UN Convention to Combat Desertification came into force in 1996 and strategies to control desertification and mitigate the effects of drought include several forestry activities that involve integrated development and management of susceptible lands. In Mali, for example, such forestry activities include afforestation, natural plant management, silvo-pastoral systems, agroforestry systems, watershed management and development of national parks (Berthe, 1997). In northern China massive 'shelterbelts' of *Populus* spp. are being used for this purpose (Carle and Ma, 2005).

To understand the role trees play it is useful to outline the crucial steps leading to desert conditions, a trend caused by poor rains but aggravated by over-grazing, trampling and compaction of ground, savanna burning, and excessive gathering of firewood and fodder.

1. Too much pressure on land leads to a loss in the complete grass/vegetation cover and some patches of soil become exposed.
2. Step 1 leads to greater pressure on remaining grass and trees, which in turn further exposes soil to wind erosion. Adjacent patches of degraded and exposed soil start to coalesce.
3. Eventually a time comes when trees and shrubs disappear along with most of the vegetation, soil erosion greatly increases and soil fertility declines rapidly. Degraded patches expand further, forming desert-like areas.
4. Extensive wind erosion takes place, producing dust and sand storms which blow up and move quickly over land now largely devoid of shelter. Any remaining vegetation only grows slowly, is under great grazing pressure, and suffers serious abrasion in storms. Ditches and irrigation channels become blocked and even whole fields submerged by blown sand or movement of sand dunes. Once fields are covered or vegetation gone, land becomes worthless, at least temporarily, and the cultivator/grazier moves to new ground and adds pressure there.
5. Desert is left.

Tree planting can halt this process and reverse the trend of environmental degradation and is an important component of better management of the entire renewable resource base (Ben Salem, 1991; Berthe, 1997). The object of such planting, e.g. the 'green belt' projects and social/community forestry programmes of many countries bordering the Sahara and China's Three North Shelterbelt Programme, is not so much to create a physical barrier to desert encroachment (which people like Richard St. Barbe Baker advocated 70 years ago), since this suggests wrongly that most desert expansion is due to moving sand dunes and land being swamped, but to bring about localized environmental improvement and provision of people's needs so as to relieve the desert-causing pressures, such as the example in Fig. 6.2(b). Belts of trees, along with introduction of agroforestry practices, as part of an overall integrated land-use strategy, can ameliorate the harsh environment and living conditions in many ways, namely, reduced soil erosion, increased supplies of firewood, provision of shade and shelter, less extreme desiccation of crops during strong winds, and trapping of sand and dust. The momentum generated in reversing the trend will itself help to stem the land-use practices that aggravate desertification. The several benefits of tree planting combine together to effect improvement, just as the initial excessive pressure leads to rapid decline towards desert; unfortunately, the restoration process through tree planting, and creation of planted forests, takes much longer and silviculturally is often challenging (Evans and Turnbull, 2004; Carle and Ma, 2005).

6.5 Ecological Roles

Several of the protective roles of planted forests have as a primary objective conservation or restoration of natural ecosystems and their germplasm. Four key roles are: protection of threatened ecosystems, rehabilitation and restoration of damaged or degraded forest, reclamation of sites, and ex-situ conservation of genetic resources. The subject is huge and of great interest to many.

6.5.1 Can planted forests protect threatened natural ecosystems?

The concept of an 'island' of undisturbed native forest protected by a collar of planted forests to act as a buffer is appealing. It suggests deliberate attempts at conservation. Such buffer zones certainly bring micro-climatic amelioration in the transition from non-forest to native forest and may even assist wildlife conservation at the forest edge (Boston and Sessions, 2006; Denyer *et al.*, 2006). A far commoner situation is fragmentation of native forest. Many authors have explored the impact that planting trees and planted forests have, For example, restoring tropical diversity by speeding up secondary succession in gaps by planting (Martinez-Garza and Howe, 2003); examining the impact on native bird species in a landscape mosaic of planted *P. radiata* forests and fragments of native Maulino forest in Chile (Vergara and Simonetti, 2006); comparing diversity and conservation of bird assemblages between mature planted pine forests and fragments of *Quercus ilex* woodlands in Spain (Santos *et al.*, 2006); and whether frog populations decline in planted forests of *P. radiata* in Australia compared with remnants of native woodland (Parris and Lindenmayer, 2004).

Parallel to these questions of island biogeography is the specific issue of ecological corridors and their value. Numerous authors have written on this topic, for example, Bollen and Donati (2006) discuss conservation of littoral forest in Madagascar and the threat planted forests pose to wildlife unless forest corridors connect isolated fragments of native forest to allow exchange of genes in endemic plant and animal populations; Poulsen (2001) describes design of large planted forests with corridors to link up retained patches of native forest; and Strauss (2001) reports how large planted forests of native *E. globulus* in Tasmania, Australia, provide a far better corridor for native species than plantations of exotic *P. radiata*.

The above studies, which are no more than illustrative, are part of a corpus of knowledge and experience that has built up mainly over the last 20 years. They sit alongside research into planted forests and biodiversity generally, into establishing them specifically for wildlife benefits (e.g. Twedt and Wilson's (2002) and Twedt *et al.*'s (2002) accounts of *Quercus* and *Populus* spp. plantings in Mississippi and the Louisiana bottomlands, USA), how to plan and design new planted forests to conserve wildlife (Zurita *et al.*, 2006), including creating new native woodland (Rodwell and Paterson, 1994), and what interventions can be made to enhance biodiversity in existing planted forests (Ferris-Kaan, 1995; Ferris and Carter, 2000).

What is clear is that using planted forests to help conserve threatened ecosystems can be significant but will never be a major role. What is emerging as a major role is conservation by default as well as by design. Application of best practice guidelines avoids damaging important ecosystems in areas of planted forest, such as wetlands, fragments of native forests, and even individual ancient trees. Because planted forests, in particular uniform stands of trees, can be unattractive, there are generally fewer human visitors compared with open land. Sight, sound and smell of humans diminish and wildlife returns: planted forests may not be ideal habitat, but absence of people is. Evans and Turnbull (2004) cite the example of the Usutu Forest, Swaziland as representative and that planted forest of exotic *P. patula* and *P. elliottii* also demonstrates another conservation benefit 'by default'. The rides, firebreaks and other unplanted areas that break up blocks of forest have become rich herb meadows and floristically diverse, even containing endangered species (Evans and Masson, 2001). Similar 'refugia' have been observed in the UK, where tracks and rides in planted forests provide remnant habitat for native grassland and meadow plant assemblages, more than 95% of which elsewhere have long since disappeared under the plough.

In all development and management of planted forests, to enhance biodiversity the key conclusions are:

- Conserve existing natural features – fragments of native forest (in their own right and as seed sources), wetlands, and other special habitats – both by avoiding planting them and even reinstating them where possible in existing planted forests.
- Retain individual ancient, unusual or very large trees.
- Encourage development of trees and stands of all ages and conditions, including extending rotation length to simulate 'old growth' conditions.
- Avoid uniformity and, in particular, diversify structure – edge habitat, open areas, tall trees, new growth.
- Provide areas of light and shade – glades, open rides, firebreaks.
- Leave some dead trees standing – snags – and create deadwood piles.
- Resume traditional practices, where appropriate, such as coppicing and pollarding.
- Minimize soil disturbance and loss of soil seedbank.
- Manage 'new' habitat such as fire dams with a view to their wetland potential.
- Choose native species for planting in preference to exotic.

6.5.2 Planting trees and forest for ecosystem rehabilitation and restoration

Globally, the damage to native forest formations has been both decline in extent (deforestation) and decline in condition (forest degradation), leading to loss of productivity in terms wood and non-wood products and of environmental services. In particular biodiversity has been greatly reduced, not only in the species-rich tropical rainforests, and few areas can recover wholly unaided. Can tree planting aid this recovery and rebuild indigenous forest ecosystems, a process known generically as 'forest restoration'? It is a role that is increasing in importance.

Forest restoration is being addressed by the United Nations Forum on Forests, the Convention on Biological Diversity, and, as noted, restoration of vegetation in arid and semi-arid areas is a key aim of the Convention to Combat Desertification. While planted forests will never restore all the products and services that native forests provide or the needs of key interest groups, they can play a vital role in restoring forest benefits at the landscape level (Maginnis and Jackson, 2003). At the local level, when rehabilitation and/or restoration are prime objectives, planted forest development must plan for diversity within the landscape and modify silvicultural and management practices. Forest restoration is considered one of the great challenges of the 21st century (Sayer, 2002) and will add further to the area classified as 'planted component of semi-natural forests' (Chapter 3).

Different approaches will be needed at different sites and the degree of forest degradation will limit what restoration goals are achievable. Planted forests can play several roles in rehabilitation and restoration of ecosystems to achieve objectives of biodiversity conservation. The first step, however, is to define terms. For example, ecosystems can be characterized by their species composition, 'structure' (i.e. complexity), and 'function' (i.e. biomass and nutrient content). If any of these are reduced, the ecosystem is degraded. So when a forest has been selectively logged and its complexity and possibly species richness reduced, it is considered 'degraded' even though its biomass and productivity may remain quite high. Three levels of increasing site disturbance have been identified by Aber (1987): (i) disruption or removal of the native plant community, without severe soil disturbance; (ii) damage to both vegetation and soil; and (iii) vegetation completely removed and the soil converted to a state outside natural conditions. Three approaches, 'restoration', 'rehabilitation' and 'reclamation', are the terms that describe reversing the degradation process and assisting recovery of these different degrees of disturbance. Figure 6.7 illustrates these issues.

In this section restoration and rehabilitation are outlined together, followed by reclamation in the next section.

Restoration

Restoration generally means 'to bring back to the original state or condition'. So, the thrust of ecosystem restoration is to try to re-establish the presumed productivity and species diversity of the forest condition at a particular site before it was disturbed. It attempts to match the ecological processes and functions of the original forest (Lamb, 2001).

One of the basic questions of restoration ecology is how to define what is meant by the 'original ecosystem'. Replacing the exact species composition of the original forest is impossible, even if this was known with any certainty. The aim is to initiate successional development where natural recovery is impossible, or accelerate it when it is feasible. Moreover, how restoration is done depends on the extent of degradation and resources available and will not necessarily replicate how the ecosystem developed under natural conditions.

Rehabilitation

Returning any converted or damaged forest lands to a functioning forest is considered 'rehabilitation' by Brown and Lugo (1994) and Lamb (2001). Lamb (2001) defines 'rehabilitation' as 'to re-establish the productivity and some, but not necessarily all, of the plant and animal species thought to be originally present at the site. For ecological or economic reasons, the new forest might also include species not originally present at the site. The protective function and many of the ecological services of the original forest may be re-established.' Rehabilitation is easier to achieve than restoration and, by including commercial species, can

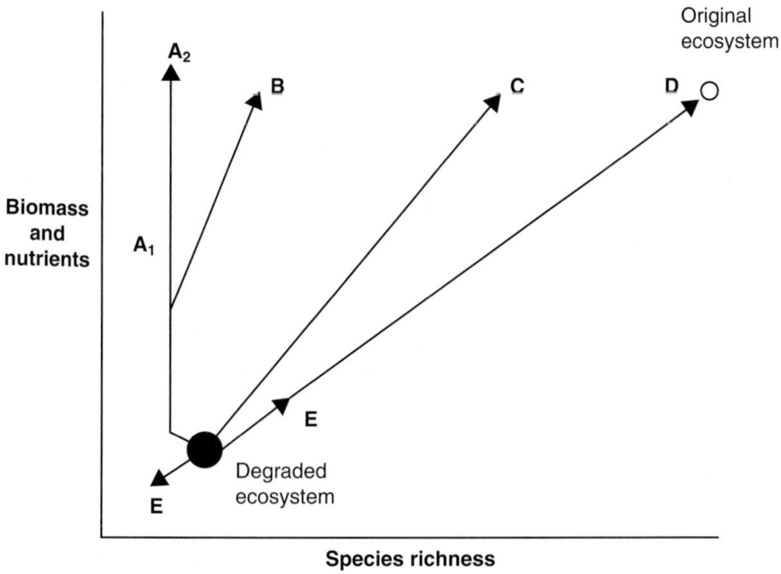

Fig. 6.7. Relationship complexity and productivity of ecosystem development for reclamation, rehabilitation and restoration. Species and complexity describe ecosystem structure, and biomass and nutrient content describe ecosystem function. *Reclamation* involves plantation monoculture of exotic species (A) which may have a biomass less than A_1 or greater than A_2, the original biomass depending on weedicide and fertilizer use. These may acquire a diverse understorey over time (B). *Rehabilitation* involves both native and exotic species (C), while *restoration* (D) leads to a new ecosystem approximating the original ecosystem. If the degraded ecosystem is neglected it may recover or degrade further (E). Based on Lamb (1994) and Bradshaw (1987).

often have a more favourable economic outcome. In fact, Lamb (1998) suggests using commercial planted forests for timber production with modified designs as a promising approach to restore on a large scale a proportion of the former biodiversity of degraded forest lands.

Issues in rehabilitation and restoration

Planting for restoration, rehabilitation and, indeed, reclamation, can be done on any individual site, but there is growing support for 'forest landscape restoration'. This process aims to regain the ecological integrity and enhance human well-being in deforested or degraded forest landscapes (Maginnis and Jackson, 2003). It is more than just adding some tree cover. It provides at the landscape level a variety of forest products, services, ecological processes that benefit both people and the environment. It has both socio-economic and ecological dimensions (WWF, 2002). There is clearly a role for planting and planted forests to complement other strategies, such as ecological corridors, conserving remnants of native forests and agroforestry systems: the role itself depends on how degraded the landscape is and what main improvements are sought.

Where residual forests remain, their protection from further degradation may be sufficient, but often enrichment by planting in lines or clearings is required. But where no native forest remains, re-introduction of the woodland environment by planting is required. Lamb (1994) outlines seven models: planted monocultures of native species, mosaics of tree monocultures across the landscape, agroforestry, multi-species planted forest, nurse trees, underplanting beneath earlier monocultures, and encouraging understorey development (Fig. 2.4). All beg the question – what happens next, when has rehabilitation finished?

Planted forests influence the local environment and may allow biodiversity to increase over time if new species can colonize the sites (Lugo *et al.*, 1993). They may modify *microclimate* through: (i) less fluctuation in temperature and humidity in the stand; (ii) better light environment for seed germination and seedling growth; and modify *soil physical and chemical properties through*: (i) reduction in erosion rates; (ii) increased soil organic matter and cation exchange capacity; and (iii) increased soil nutrient flow through litter decomposition and soil microbiological activities.

These changes stop or slow the degradation process and enable less hardy native plants species to regenerate in the understorey (e.g. Lamb and Tomlinson, 1994; Fimbel and Fimbel, 1996; Huttel and Loumeto, 2001; Carnevale and Montagnini, 2002). This is accompanied by changes in vertebrate fauna (birds, small mammals etc.) and soil macrofauna (earthworms, termites etc.) (e.g. Bernard-Reversat, 2001).

The primary objective for restoration planting will be ecological in nature and aim to re-establish the original plant and animal communities to a site in a particular time frame and within the resources available. The first stage may be ecologically crude, such as simply establishing some sort of permanent cover (Bradshaw, 1987). But even this requires ecological knowledge and use of appropriate practices and, importantly, discussions on socio-economic issues with stakeholders. Participatory and adaptive management planning is essential to determine what strategies best achieve restoration or rehabilitation.

Where the objective is primarily rehabilitation, the potential for an economic return is addressed alongside increasing the structural complexity and biodiversity of the planted forest. And it is no less important to address local social issues with stakeholders. For example closing areas for rehabilitation or restoration may impact people's livelihoods if the site is used for grazing animals, collection of fuel and non-wood forest products, or gives access to other areas. Landholders may assist with rehabilitation if there is some economic return from otherwise unproductive land, but may be quite indifferent or even antagonistic if no benefit is perceived. This was the case in a dry region of Yunnan province in China, where villagers grazed their goats on highly eroded and virtually treeless slopes until they had the opportunity to plant neem, *Azadirachta indica*, to produce a saleable product (Liu, 2003). Without the incentive of recouping their planted forests' costs and making a profit, they had little interest in rehabilitating the site.

In purely restoration plantings where the soil is left largely intact, as in some recently logged rainforests, planting native species chosen from nearby undisturbed communities may be feasible (e.g. Mori *et al.*, 2000). Then, provided the forest is not too fragmented, succession may occur naturally with seed from adjacent native forest. However, if most native forest is gone, such as the abandoned pastures in Amazonia, planting 'tree islands' of species that grow rapidly in full sunlight and produce fleshy fruits soon after planting to attract seed-carrying birds and bats has been recommended (Nepstad *et al.*, 1991). Similarly, ecological rehabilitation by planting 'framework species' can accelerate succession (Tucker and Murphy, 1997). About 20 native species are selected that are able to shade out weeds and attract seed-dispersing wildlife. The mixture of framework trees includes fast-growing pioneers able to occupy the site quickly and slower growing, large-seeded species that are less easily dispersed (Forest Restoration Research Unit, 2000).

For biodiversity conservation and restoration planting, native species are strongly preferred or essential. In the tropics this raises a challenging issue that for many native species little is known of their physiological tolerances, silvicultural characteristics and best means of propagation. More attention is being paid to such studies, e.g. in Brazil (Knowles and Parrotta, 1995), Costa Rica (Butterfield, 1995) and Thailand (Hardwick *et al.*, 1997; Forest Restoration Research Unit, 2000). From a practical point of view, species that seed regularly can be grown easily in the nursery, and that are tolerant of a range of site conditions are the most useful, particularly if already well known, e.g. *Cordia alliodora* in the neotropics (Fig. 2.4).

Although exotic species may not be very attractive to native wildlife (Wunderle, 1997), they do modify site conditions, bring back the forest environment and with appropriate management can increase biodiversity in the understorey and soil (Fimbel and Fimbel, 1996; Geldenhuys, 1997; Bernard-Reversat, 2001). Nitrogen-fixing hardwood species, such as *Acacia auriculiformis*, *Paraserianthes falcataria* and *Alnus* spp., may improve soil nutrient levels more rapidly than non-nitrogen-fixing stands through higher rates of nutrient cycling and facilitate development of native understorey vegetation. However, one risk with exotic species is that of their becoming invasive. The removal of such unwanted planted trees from native woodland sites or the landscape generally is an issue that is expected to increase.

6.5.3 Reclamation of special kinds of sites

This last role of tree planting is to use trees as agents of rehabilitation of a site and refers to revegetating highly degraded sites, such as mined or salt-affected lands. It aims 'to recover productivity of a degraded site mostly using exotic tree species', but 'The original biodiversity is not recovered although the protective function and many of the ecological services may be re-established' (Lamb, 2001). Species with exceptional physiological tolerances to improve site conditions and initiate soil-forming processes means that species of *Acacia*, *Alnus*, *Betula*, *Eucalyptus*, *Pinus*, *Salix* and other pioneers are frequently employed.

Trees planted in reclamation serve several purposes:

1. Stabilize soil by preventing the surface from moving as root systems develop and ramify in the substrate.

2. Protect bare ground from erosive forces provided grazing is prevented and ground vegetation is encouraged to develop.
3. Begin the gradual process of soil development through litter fall and the build up of organic matter, soil organisms and microflora. Nitrogen-fixing species are particularly useful since many tolerate harsh environments and are able to grow on nitrogen-deficient substrates typical of industrial wastes. Also, their foliage, which is rich in nitrogen and base nutrients, is usually readily decayed and incorporated.
4. Overcome problems of contaminated land through 'phytoremediation', for which *Salix* species can be especially valuable (Dimitriou and Aronsson, 2005).
5. Restore a woodland habitat and the related wildlife that usually returns.

Sites for reclamation are planted because little else will grow readily and often because of the need to stabilize the ground surface, the presence of toxic substances or extreme infertility. Examples of species include *Casuarina equisetifolia* in China, India, Senegal and Vietnam (Midgley *et al.*, 1983; Lam, 1998) and *E. camaldulensis* and *Acacia ampliceps* to reclaim saline and sodic sites (Marcar and Khanna, 1997). In the tropics, afforestation of industrial wastes include planting *Acacia auriculiformis* on copper tailings in the Philippines and tin tailings in Malaysia, *E. tereticornis* on copper tailings in Papua New Guinea, *P. caribaea* var. *hondurensis* on restored opencast iron/nickel workings in the Dominican Republic, and *E. camaldulensis* on waste land from opencast tin mining on the Jos plateau in Nigeria (Evans and Turnbull, 2004). In temperate countries afforestation to reclaim mining lands, such as after bauxite mining in Australia (Gardner, 2001), strip mining in the USA (Ashby and Vogel, 1993), or open caste mining in the UK (Moffat, 2004) is widespread. The focus of such reclamation is as much on insisting that topsoil, or at least suitable soil-forming materials, are retained during site restoration and then redistributed over the surface, and that compaction is avoided, as on choice of tolerant tree species (Moffat and McNeill, 1994; Savill *et al.*, 1997; Moffat, 2004).

6.5.4 Planted forests and conservation of genetic resources

The direct role planted forests play in conservation of genetic resources is limited, and is rarely if ever a principal object of management. The obvious exceptions are arboreta and similar great tree collections noted earlier. However, countless seed collections, followed by tree introductions, species trials and tree breeding programmes at the heart of planted forest silviculture, have created banks of germplasm which, today, may sometimes be of genuine importance for ex-situ conservation of genetic resources. This arises from loss of native forest where the species occurs, e.g. many tropical *Pinus* areas in Central America, or where seed collections in remote areas coincided with good fruiting of otherwise difficult to obtain species.

There are several species where the area of planted forests now far exceeds their extent in their native habitat. The best example is *P. radiata*, but it is also true of eucalypts such as *E. grandis* and *E. urophylla*, as well as some other species.

7 Policy, Institutional and Ownership Issues

D.A. NEILSON AND J. EVANS

7.1 Introduction

Several themes emerge from earlier chapters that suggest change in the policy, legal and institutional frameworks. Who will own planted forests? How will they be funded? What roles will different stakeholders play? Will certification effectively become the sector's global policeman? And will illegal practices, so rife in much of tropical forestry, also impact planted forests and muddy the outlook? There are many other issues – some of the social ones were discussed in the previous chapter – but the ones here are addressed as critical to the world's planted forest estate and the future it offers. Compared with most native forest, the issues are less intractable since planted forests usually have distinct boundaries, will have required investment to create and to care for them, and are managed in accordance with some specific objective(s): their artificial nature makes for less complexity, as challenging as some of the issues still are! These characteristics are well recognized (Garforth *et al.*, 2005a) and can assist, for example, where state-owned resources are transferred to the community or the private sector.

7.2 Who Owns the World's Planted Forests?

In Chapter 2 the trend in ownership from public to private in the last 30 years was noted, and in FAO's thematic study of planted forests this change was explored further. Chapter 4 reports data showing that for planted forests for production, public ownership fell from 70% to 50% in the last 15 years, but where the protective function was uppermost, the bulk remained in public ownership. And, for both function types, there is an additional trend to increased smallholder ownership. However, the bigger questions remain: are these trends, based on limited data, representative of the whole planted forests estate and will they continue into the future? Also, drawing on Neilson (2007a), evidence from corporate investment and private-sector funding is examined in detail.

In *Policy that Works for Forests and People*, Mayers and Bass (2004) point out that companies that take the longer view in their investment decisions are moving towards planted forests because of practical, logistical and tenure advantages. Indeed, they go on to say that most of the world's high-yielding forests are owned by corporations with access to the technologies that can support such production. But ownership isn't simply about property rights to land and forest; as noted later funds are becoming more mobile with investors' time-horizons often very short by forestry standards. Also, as governments – in effect the public sector – divest themselves of physical assets and resources, the process is often in stages – outsourcing, transferring of rights and ultimately of ownership itself. Garforth *et al.* (2005a), in examining this evolution, also introduce the special case of restitution, where previously owned assets that were seized by the state are returned or restored. They cite an estimated 3–4 million new private forest holdings, albeit owning only a few hectares each, across the ten accession countries joining the EU in 2004. For Romania, Ioras and Abrudan (2006) report that some 2 million hectares have been legally claimed, but restitution is limited to 10 ha per individual, 20 ha per community individual, i.e. forests previously owned by towns and villages, and 30 ha for churches and schools. It underscores the trend of increased smallholder ownership as well as that of decreased ownership by the public sector.

The climate of decentralization suggests these trends will continue. There will always be a continuum of privatization models, but increasingly governments will confine their role to setting an enabling and regulatory framework through an array of fiscal, policy and legal instruments, rather than resource owners and managers. How much the distinction between production objectives for planted forests, which generate direct economic returns, and protection objectives differentiates future ownership is unclear. Traditionally the state is seen as custodian of land and landscape managed for the common good – national parks and reserves for protecting the environment and important ecosystems, and the private sector for commercial forestry activity, but this is not essential. Legal requirements on the private sector to deliver such non-market benefits, through felling regulation, structured incentives, silvicultural guidelines and codes of practice, allow the state to deliver its duty as custodian without ownership, a route partially followed in the UK and wholly so in New Zealand.

The long-held view that state involvement in forestry, including planted forests, alongside the private sector, enabled the regulatory environment to be better informed through on-site experience of costs, returns, markets etc. remains true, but is not considered essential today. It does not apply in agriculture, and though the long timescales in forestry are a special case, there are other means to garner the information than being a participant. It also removes the contradiction, or potential conflict of interest, of government as regulator and as manager.

A more uncertain question concerns afforestation for carbon sequestration. Will the small owner benefit? Are massive carbon-offset plantings best managed by para-statal organizations or non-profit bodies to ensure their permanence and compliance? Or, with appropriate controls in place, will carbon accounting become another income stream for any planted forest regardless of ownership?

Planted forests appear increasingly to be the domain of the private sector, both the corporate on the one hand and the private smallholder on the other, with the state playing a diminishing role except for providing and policing the regulatory and enabling environment. Consequently supply of forest products will increasingly be in private hands, with the state retaining ownership only of areas of high conservation, heritage or recreational value. Even where the state has other reasons for retaining ownership of the land, e.g. political or on behalf indigenous peoples, rights are often assigned to the private sector to establish and manage planted forests.

Before considering the major funding and investment possibilities and consequences of this trend in ownership, further comment is made on the increase in smallholder ownership.

7.2.1 The issue of scale for the small owner

The proliferation of smallholder ownership can bring problems of access to market intelligence, poor technical knowledge and support, inefficiency in operations – or simply poorer management (Ioras and Abrudan, 2006), as well as increased bureaucracy for state administration (see Box 7.1; from Carle, 2007). Traditionally extension services, common-interest groups and cooperatives have responded to such weaknesses. In principle, the issues are no different from smallholder farming, but longer rotations and the usually smaller returns from planting and growing trees limit what solutions work. While some success

Fig. 7.1. A significant harvest for the owner of this 12 ha planted woodlot, but a more complex contribution to the overall wood supply market.

Box 7.1. Vulnerabilities of smallholder plantings

A significant portion of planted forests (including forest plantations and the planted component of semi-natural forests) are owned by smallholders. Of the 272 million hectares of planted forests globally (excluding trees outside forests), smallholders (less than 100 ha, greater than 0.5 ha) own 26%, the corporate private sector 15% and governments 59% (FAO, 2006a). Moreover, of the 205 million hectares of planted forests managed for productive functions, smallholders own 32%, the corporate private sector 18% and governments 50%.

Smallholder planted forests are characterized by diverse sites, mechanisms, species, silvicultural practices, rotations and end uses tailored to suit smallholders' own livelihood needs and support sustainable land use. In developing countries smallholder plantings can contribute positively to achieving food security and alleviating poverty. For these reasons governments and the corporate private sector can establish policies and incentives that encourage smallholders to invest in planted forests. However, many smallholders, particularly in developing countries, face unique commercial, technical, policy, legal, regulatory and institutional vulnerabilities that need to be recognized and overcome. These may include:

- Lack of clear enabling policies and procedures regarding land tenure, crop ownership and rights to manage, harvest, transport and market forest products.
- Complex incentives and insufficient access to development funds to invest in planted forests, exacerbated by high interest rates and the stringent requirements for collateral against funds borrowed.
- Weak technical knowledge and poor access to information on germplasm and reproductive materials, nursery practices, site preparation, planting, tending, silviculture, protection (against insects, diseases, other pests, fire), harvesting and the measurement of volume and value of harvested forest products.
- Weak commercial and business knowledge and weak bargaining power in contracts and agreements (harvesting, transport, sales).
- Limited access to market information on products, specifications, market prices, harvesting and transporting contract rates and the implications for investment and management decision-making.
- Fragmentation and isolation of small production units, and attendant problems of basic communications and networking.
- Insufficient resources of government technical and extension services to support smallholder investment in planted forests.
- Disadvantages that management tools for sustainable forest management (e.g. codes of best practice, certification) can represent for smallholder investment.

These vulnerabilities can increase risk and result in suboptimal technical and commercial performance of smallholder investments in planted forests – poor quality of planted forests and forest products, low yields, high contract prices, low market prices and inequitable sharing of benefits.

Smallholder associations, in addition to governments and NGOs, have an increasingly important role in representing their members in policy dialogue, providing technical support and extension and assisting marketing and business decision making to improve the smallholder returns on investment in planted forests.

has been had, e.g. in South Africa or non-industrial private forest in the USA, and in small island states such as the Solomon Islands, smallholder plantings can be a way of bridging a resource gap (Raymond and Wooff, 2006), the outlook for development of small planted forests as a critical component of the planted forest resource is not yet mature in market terms (Fig. 7.1) as well as the issues listed in Box 7.1. If failure to mature this sector is widespread, there is the risk, noted in Chapter 4, of wholesale switching out of tree-growing into another land use with the supply consequences this entails.

Of course, mobilizing many small-scale plantings has long been in the suite of national forest policy objectives. But these are often for non-market benefits, either centrally inspired such as China's massive village-based/led planting campaigns for soil protection, or with incentives to deliver environmental benefits, such as the current England Forestry Strategy, rather than economic returns.

7.3 Funding and Corporate Private Sector Investment

7.3.1 The role of the market

Most planted forests around the world were initially established by governments. This was natural as governments owned most original forestland, and so were charged with managing these forests after harvesting (which in many instances meant planting trees) or played active 'national good' roles in establishing new planted forests, e.g. the UK, South Africa, Australia, New Zealand and Chile. These public investments in planted forests by governments were often production-driven to provide future wood supplies for processing expansion and/or to demonstrate methods for planted forests' management.

Although private ownership of planted forests has a long history, it became increasingly significant from the 1950s as global forest products companies expanded planted forest areas to secure wood resources for future processing demand. This production-driven mentality continued until the late 20th century, but since then the private sector has responded increasingly to international and national market and commercial opportunities to invest in planted forests purely for financial returns.

At much the same time some governments decided to dispose of all or part of their plantings to private owners. These included the governments of Chile, New Zealand, South Africa, Portugal and some Australian states. In most cases governments continued to own the land, but sold the trees and rights to grow future rotations on that land. Usually the land itself was not sold because of possible complications relating to indigenous peoples' rights. In addition some governments, including Indonesia, Malaysia, China and Russia, have provided long-term land concessions to companies in order to grow trees.

There are, nevertheless, major parts of the world, such as Central and Eastern Europe, Western Asia and Africa, where government ownership of forest still dominates the industry. This situation may change in the future as increasingly democratic governments realize that owning forests is not essential for delivery of policies in the national interest.

7.3.2 Recent changes in investment and corporate ownership of planted forests

Of significance, however, is that only recently has substantial ownership of planted forests moved from a production-driven motivation (securing wood supplies for processing) to more commercially motivated reasons. As a result, ownership patterns have changed and continue to change.

The general relaxation of investment and trade restrictions leading to globalization has included international investment in planted forests. Since the 1980s ownership of planted forests around the world by institutional pension and endowment funds has risen significantly. This began in the USA, with domestic institutions diversifying their investments into categories such as North American planted and managed native forests. By 2007 this region had the highest investment in planted and managed native forests owned by institutional funds, although there were other smaller funds based in Europe, Oceania, South Africa and Latin America.

The scale of the ownership change from forest products companies to institutional and other owners has been phenomenal. In the USA in 1981 forest products companies owned 23.5 million hectares of managed forests, by 2007 this figure fell to below 6 million hectares. Over much the same period investment by institutional funds in global planted and managed native forests rose 30-fold, from less than US$1 billion to an estimated US$30 billion.

This changing ownership has given rise to several new investment vehicles where investor and manager of the resource may be quite separate. Often institutional investors do not want to buy and manage planted forests, or may only want to invest small amounts of their funds, and there is a need for a structure that can accumulate funds from many organizations to provide large funds. Other approaches include tax efficient, publicly listed investment trusts. Indeed, the largest private planted forest owner in the world (Plum Creek, USA) is one such and, in countries with the appropriate fiscal environment, this approach is expected to increase rapidly both for new investors and by restructuring existing forest products companies. And, most recently, yet another class of forest owner has emerged – the huge 'hedge funds' that may buy and sell, or buy and hold vast tracts of planted and managed native forests. Often they would immediately set out to break them up into smaller parcels to re-sell, with interesting implications for management and sustainability.

Since 2005 a number of European-based private and listed funds have begun investing in planted forests in both Europe and elsewhere, such as the USA and Australia. Some aim specifically to invest where potential carbon trading rights are an attraction. In addition there is the new phenomenon of specialist country funds. For instance, in 2007 a Colombian planted forest investment fund was launched, and a similar Japanese planted forest fund is being considered for launch in 2008/09. Yet another source of funds that may play a significant role in planted forest investment is the rapidly growing sector called Sovereign Wealth Funds (SWF) – national investment authorities, state superannuation funds etc.

Many of these new investment vehicles restrict investment to planted forests, shying away from native forests with all their attendant environmental and social issues. Also they focus on countries where secure land tenure, strong government institutions and judicial independence exist. Another feature is that they rarely invest in planted forest expansion. They may replant after harvesting, though some may not even do that, but only a few in very selected regions seek new land to establish an increasing planted forest area. This may change as the opportunity to invest in existing forests reduces.

Not all forays into investing in planted forests have succeeded, e.g. the misguided teak projects in Central America and India in the 1980s and 1990s. However, even these, along with other native hardwood replacements, are being re-evaluated. Escalating land prices, for example in Latin America, may limit teak expansion, but planted forest investors are being encouraged to consider hardwood investments in higher risk countries.

There are exceptions to the above – that forest products companies sell their planted forests to institutional funds and rely instead on long-term contracts for wood supplies – notably Japanese pulp and paper companies. They have invested directly in fast-growing, mainly hardwood crops in countries such as Australia, Chile, Ecuador, Brazil, South Africa, Laos and China to secure fibre supplies, though they are increasingly running up against competition from the USA pension and endowment funds in seeking to secure land and planted forests.

7.3.3 Mobility of investments

International funds expect a competitive return on their investments within acceptable levels of risk. If the risk becomes unacceptable or the returns are insufficient, the new investors in planted forest can be expected to sell and invest elsewhere. For this reason, while planted forests investments themselves are very immobile (being locked into the land that is planted), planted forest investment funds by contrast have become very mobile. It is now common to buy planted forests (and managed native forests), and then sell on part or all within a decade. Endowment and hedge funds may buy and sell on very short timeframes. For example, recently The Harvard Endowment Fund bought and then re-sold over 300,000 hectares of forestland, mostly in the US South, over an 18–24-month period, and again in the USA a hedge fund bought all of Boise Cascade's planted forests, parcelled them and sold them on within 12–18 months.

Planted forest is increasingly viewed like any other investment (Fig. 7.2). That said, this 'short-termism' of the forest owners may be creating a new set of problems, including fire management and long-term wood supply security (or lack of it) for wood-processing companies.

7.3.4 Incentives – subsidies

There has been one very important factor that has linked almost all successful planted forest expansion projects internationally: the application by governments of generous subsidies or tax concessions. There is always criticism of these 'free

Fig. 7.2. Planted forest of *Pinus radiata* in New Zealand – today, just another investment? (Source: FAO.)

handouts' and that the schemes attract 'fast money' investors motivated by greed and not the worthiness of the project. In some cases it has led to planted forests being established in the wrong areas, outside sensible guidelines for suitable soils, rainfall and other factors fundamental to success. The reality, however, is that without generous payments and concession schemes, major planted forest expansion projects rarely, if ever, get off the ground.

There are several outstanding examples.

- Chilean Decree Law 701, in which the government underwrote most of the development and management costs of establishing planted forests. The massive expansion has led to the Chilean forest products industry becoming a world giant.
- Generous tax-based schemes in Brazil from 1967 to the mid-1980s, which significantly expanded the softwood and eucalypt planted forest base (Evans and Turnbull, 2004). The expansion stopped almost immediately the scheme was discontinued and Brazil now suffers from a planted forest softwood shortage as a result.
- In Uruguay a focused subsidy scheme successfully increased the area of planted forests and the country remains very attractive for overseas investors.
- A change to the tax structure in New Zealand in 1992 enabled a major expansion of its planted forest resources for the next decade.
- Attractive subsidies and tax concessions enabled Indonesian pulp companies to establish several hundred thousand hectares of fast-growing planted forests of *Eucalyptus* and *Acacia* crops. Unfortunately the subsidies caused massive clearance of native forest, though they did largely achieve the objective of developing a planted forest industry, and lax monitoring allowed

some entrepreneurs to claim benefits without establishing or properly managing the trees.
- Generous subsidies and land rental deferment policies in Vietnam have enabled a very large planted forest base to be established, mostly by small farmers.
- One of the most attractive tax concession schemes for expanding planted forests is offered by the Australian government and has resulted in several hundred thousand hectares being established since the late 1990s. Most of the money has come from small, independent investors, grouped into syndicates under 'Managed Investment Scheme' projects. Demand has been so great that the price of rural land tripled in 5 years, and has caused serious competitive issues with farmers.

However, it needs noting that schemes that have worked in the past can later falter, as noted above for Brazil. Another example is the grants for tree planting and tax incentives provided by the UK government. These began in the late 1950s and, for over three decades, led to a major expansion of commercial conifer (softwood) planted forests. Today, while many of these incentives are still in place, they are now redirected to less commercially attractive native hardwood species. Not surprisingly, expansion of new planted forest has more than halved and the new planting of conifers has fallen to a tenth of what it once was (Forestry Commission, 2007). It is an example of a planted forest incentive policy basically failing to attract significant investors – or perceived as having largely achieved its aims – because its basic philosophy has switched from what would attract planted forest investors to that of paying for what taxpayers want – better public access, beautiful landscapes and wildlife conservation in woodlands.

In summary, the positive impact of generous subsidies and tax concessions can be seen in a great many examples around the world. In contrast, investor behaviour in not planting trees – when there are no subsidies or when subsidies are removed – provides a very clear picture of the critical importance of incentives to global planted forest development.

7.3.5 Risk assessment for investors in planted forests

In considering the outlook for planted forests, the dilemmas faced by investors are significant in choosing options.

First, it is common for the safest, least risky investment, e.g. a standard species/regime in a stable Western democracy, to attract only modest returns compared with those from countries, regions, species or management regimes deemed to be 'more risky'. Such investment decisions are always a trade-off, but tools are becoming available to clarify the issues. One example is *The Global Tree Farm and Managed Forest Industry Review 2007* (Neilson, 2007b), which provides matrices of 'Internal Rate of Return' versus 'Country Attractiveness Ratings' for almost 100 'case studies' of softwood sawlog, hardwood sawlog and pulpwood investment options around the world.

Second, there are a growing number of well-established independent consulting firms that provide advice on investment proposals almost anywhere in the world. Many have decades of experience advising on valuations and due diligence of the full range of planted forest estates in both temperate and tropical countries.

Third, investors will have different risk profiles, and the country base of investors will affect their appetite and perception of risk. For instance, a Chilean investor may assess Argentina to be of less risk than would a USA investor, or an Indian investor that of Cambodia or Laos compared with a German investor. By examining international transactions of planted forests around the world it is possible to determine risk profiles of various investors.

Of course, investment is an imperfect science and it is very difficult to assess comprehensively values of planted forests. The long-term nature of even a single rotation, the large amount of international trade of wood products from most large planted forest countries, and decisions on such basic factors as a discount rate, future currency exchange rate movements and future log price assumptions all seriously impact a valuation. Assessment of future non-wood values further complicates the process.

In 2007 the Boston-based RISI organization[1] completed a major review of global tree farm investment attractiveness.[2] Using a proprietary template of 13 risk parameters, such as inflation, labour costs, corruption, strength of judiciary, land availability, access to markets etc., it determined an attractiveness score for 70 countries from 1.0 (extremely risky) to 7.0 (no risk). The least risky country was the USA, and other highly rated 'attractive' countries and regions included Scandinavia, Western Europe, Oceania and Chile. Medium-risk countries included Argentina, China, Bulgaria and Vietnam, while risky countries included Uganda, Cambodia, Madagascar, Angola and Zimbabwe.

From these analyses it was clear that high scores for policy consistency and an independent judiciary are requirements for a high rating for the attractiveness of a planted forest investment when ranked against competitor countries. Of course, these risk ratings reflect the international corporate risk profile. Local smallholder investors may spread their risk with other activities (e.g. agroforestry), have lower expectation of returns, and value more highly non-market benefits of their planted forests, leading to very different perceptions of risk.

Risk assessment from government and society perspectives

Recent experience indicates that some governments and NGOs take a hard and negative attitude towards expansion of large-scale commercial planted forests, especially in developing countries. The ogres of monoculture, indigenous people displacement, loss of biodiversity, diminished enjoyment of non-wood benefits by villagers, and even potential loss of water, soil erosion and carbon issues are all raised as negatives. Consequently, in a number of countries sound planted forest expansion programmes have floundered. This is strange, as many such countries tolerate, almost without questioning, massive expansion in agri-business crops such as palm oil, the expansion of numbers of highly polluting farm animals and rapid urbanization, all of which are more harmful to the environment and to indigenous people than almost any planted forest.

Nevertheless, in spite of environmental and social NGO criticism, for all major planted forest expansion programmes which have been responsibly managed, governments and citizens mostly look back on them decades later with a positive view.

7.3.6 The need for clear, consistent, concise strategies to support planted forests

There are numerous examples around the world that illustrate the positive effect that clear government strategies and standards can have on the expansion of planted forests; and, conversely, several examples illustrating where a lack of these can hinder expansion. As already noted, there are also situations where governments have changed policies and planted forest expansion has greatly diminished, as in the UK, or all but ceased, as in South Africa owing to conflict with water resources, and New Zealand where the private sector's carbon credits were nationalized.

Ironically, it is often over-zealous and top-down intervention by governments and NGOs that have resulted in disappointing or failed planted forest results, especially with 'Community Forest Enterprises'. The International Tropical Timber Organization identifies major challenges in avoiding these enterprises from being sidelined (ITTO, 2007). It provides examples of the huge gap between 'official' plans and actual control by villagers. Examples include:

- In Gambia, 170,000 hectares have been categorized as community forests, but only 13,000 hectares are actually in the hands of local villagers.
- In Cameroon, 4 million hectares are designated for communities, yet only 40,000 hectares (1%) are approved for legal use.

Unfortunately, well-meaning NGOs claim that people, particularly in richer countries, are needed to start paying for global community forestry enterprises, but it is difficult to see what incentive they will have. It requires political and social stability, and governments setting standards and then letting the market act. For example, huge government and NGO investments in planted forests in the Ivory Coast and Cameroon have been very disappointing, and, similarly in Vietnam, it was only when heavy government intervention and well-meaning NGO support of planted forest projects largely ceased that sustained and commercially successful planted forest projects – mainly by small landowners – began to thrive.

It is not difficult for governments to create supportive policies. They need to be clear and concise, and be left basically untouched for at least a decade to engender confidence that money invested will not be wasted or misappropriated.

7.3.7 Economic valuation of wood and other non-wood-based products

For decades the dominant driver for planted forest developments has been wood production. There are sometimes variations, for example, some investors have purchased a planted forest valued by the owners as one wood quality-type (e.g. pulpwood), but for which the investor identifies a more valuable one (e.g. sawlogs or ply-logs). Or, an investor may change the management regime to produce more high-value products than the vendor had identified – Evans and Turnbull (2004) cite several examples. But in all of these variations, the product being valued is simply the wood.

Traditionally, an investor might ascribe a lower discount rate where a long-term wood supply contract was in place because it made the investment more secure and reduced the risk of relying on the open market. However, recent purchasers of major planted forest assets have sometimes ascribed a negative value to long-term wood supply obligations, preferring to be unencumbered so they can take account of market forces.

Similarly, the traditional valuation of planted forests on freehold (fee simple) land assumed that the cost of land was a positive cost at the start of the investment cycle, and then counted as negative, but with the same real value, at the end of the investment cycle. It assumes that the land would not increase in value over time, and so was included in any valuation as a holding cost only. Non-market values, such as social and environmental benefits that the planted forests may provide (carbon sinks, source of employment, protection of soils and water, amenity and recreational values) have not in the past played a major role in valuation. However, if society values these sufficiently to pay for these services, this could change in the future.

7.3.8 Higher and Better Land Use (HBU)

In the last decade, and especially in the last 5 years, however, planted forest investors who also own the underlying land have found that much of the inherent value of the 'planted forest' is actually in the land itself. The term 'Higher or Better Land Use' (HBU) has been coined in the USA, as millions of hectares of planted (and managed native) forest land has been reassessed during and after purchase concerning its worth in an alternative land use. Large tracts of forestland may be subdivided and sold as smaller units. This 'wholesale to retail' process generally allows investors to attract higher bids for smaller blocks, even if the land use (planted forest) remains the same. In addition, part or even all of a planted forest may be subdivided and sold off as hunting and recreational blocks, or for residential or industrial subdivisions. The term HBU reflects a higher market value of the land and a better return on investment, but does not necessarily reflect more responsible land use.

This process started in the USA in the 1980s, when the St. Joe Pulp Company decided it could add value to its planted forests by shutting its pulp mill and breaking up its vast planted forest area in Florida into non-wood uses: residential subdivisions, airports, towns and so on. This has been a very successful strategy, and more and more owners are seizing this opportunity to add value not to the trees, but to the land. A recent sale of 160,000 hectares by Weyerhaeuser in the state of Georgia in the USA was broken up approximately equally into 'timberland'[3] and 'HBU' land before it was sold, and was marketed to different groups of people.

The identification that the land value may actually be the controlling influence has spread to other countries. In Latvia international investors have added greatly to the value of managed forest investments by selecting areas close to towns that may be re-zoned or have rock-gravel quarries on them which have not been recognized. In New Zealand one investor recently purchased planted forests and immediately commenced converting the land to dairy farms. By changing land use the investor increased the value several fold. And, in the UK, where there is a strong presumption against change of land use from forest, a market has emerged for very small woodlands of a few hectares for amenity and conservation which is being satisfied by investors buying large blocks and subdividing them. Price per hectare of planted forest has risen two- or three-fold in the last 5 years and bears little relation to the timber value.

7.3.9 Carbon trading and the impact on planted forest values

There has been much 'hype' about the use of trees in carbon emission abatement schemes in the last decade. The potential and, in rare occurrences, the opportunity to utilize carbon credits produced by growing trees derives from the 1992 United Nations Framework Convention on Climate Change. Originally forests were to be excluded from any protocols, but a small group of planted forest experts persuaded the Convention to include them, as they might provide carbon sequestration in future.

However, implementing the notion has proved fraught. This arises because few people really understand the dynamics of forests (including planted ones) in relation to the carbon cycle. There are many complexities and the number of carbon-based planted forest projects that have been accredited makes for lean pickings. However, despite the myriad of problems, the possibility of carbon credits and future carbon trading values is attracting increasing numbers of investment consultants and investors.

Australia has pioneered carbon trading 'contracts' based on planted forests. Contracts are either between two government departments in the same State, e.g. a government-owned forest and a government-owned power station, or offered as an 'option' to purchase carbon at a later date. Some Japanese pulp and power companies have paid 'deposits' for this right. New South Wales was the first government in the world to legislate to enable planted forest owners to separate legally the land, the trees and the carbon, so that all these 'assets' can be owned by different entities.

In 2006, the New Zealand government sanctioned what it has described as the first carbon project in the world so sanctioned which is Kyoto-compliant. This project was developed under the 'Permanent Forest Initiative', which allows the carbon ownership and trading of a forest established and not harvested for at least 35 years, and then thereafter always leaving a permanent canopy. European institutional money may fund the project.

In 2007 a number of companies listed on the 'AIM' stock exchange in London to attract international funds for planted forest investments involving the added benefit of future carbon trading. However, early indications of this trading right attracting new investment to expand planted forest areas have been modest. This may be because the carbon sequestration value of preserving existing native forests through reduced deforestation and degradation is probably more beneficial, as noted in Chapter 6. However, until values are given to carbon and trading becomes widespread, the potential of investment in planted forests for this purpose is uncertain. It must await a government or organization actually deciding to purchase forests to preserve them for their carbon.

7.3.10 The impacts of bioenergy production and planted forest development

The bioenergy revolution, which started in Austria and Sweden in the early part of this decade, is now rapidly spreading around the world, although it is still largely focused in Europe and North America. The major raw material for wood-based power generation is wood pellets. In late 2006, 288 wood pellet plants were operating in Europe, up from only 236 plants in 2005, and 80 pellet plants were operating in the USA, with another 28 expected in 2007.[4] Biofuel is being traded across continents with, for instance, biofuel shipments being made from Latin America and North America to Europe, and it is believed to Japan.

Several governments have decided to promote the use of ethanol as a substitute for gasoline. Brazil has been the pioneer, mostly using sugar cane as the raw material. The EU has a target for biofuels to make up 5.75% of all transport fuels by 2010. The USA is massively expanding bio-ethanol production, but mostly derived from maize. Over the next 10 years global production of ethanol and biodiesel are likely to increase three- to five-fold. As noted in Chapter 6, there are significant technical hurdles to be overcome before wood can be used in a cost-effective way to manufacture either ethanol or biodiesel, and major research into second-generation biofuels such as viable ligno-cellulosic ethanol production systems are underway in several countries.

To date, however, there does not appear to be any specialist wood-based planted forest expansions specifically designed to meet this increasing demand. Waste wood from forests and wood-processing plants is being used, including sawdust and woodchips. Recently major areas of wind-, insect- or fire-damaged forests in Europe, Sweden and the USA are being targeted to supply wood raw material for bioenergy use as well as short-rotation coppice biofuel crops of *Populus*, *Salix* (Fig. 6.3) and *Eucalyptus*. Even trade in this commodity is beginning, e.g. projected roundwood harvested from *Eucalyptus*-planted forest in the Republic of Congo destined for European power stations as biofuel.

One species of considerable potential for biofuel production is *Jatropha curcas*. It is a drought-resistant, inedible oilseed-bearing tree that is tolerant of poor soils unsuitable for agriculture. The Government of India has singled it out for large-scale planting with subsidies and soft-loans available to individuals and companies. The biodiesel content of *J. curcas* seeds is 35%, the tree lives for up to 50 years and can bear seeds three times per year (van Eijck and Romijn, 2008). In 2007, British Petroleum and D1 Oils announced a joint venture to accelerate the planting of *Jatropha*. The planned investment is about US$160 million over the next 5 years. D1 Oils will contribute 172,000 ha of existing planted forests in India, Southern Africa and South-east Asia, and the joint venture will have exclusive access to the elite *Jatropha* seedlings produced through D1 Oils' plant science programme. In total, some 1 million ha will be planted over the next 4 years, with an estimated 300,000 ha per year thereafter. If fully implemented, the project will become the world's largest tree-planting programme.

7.3.11 Overall outlook in planted forests investment

Massive 'investments' have been made in planted and managed native forests by institutional and high net wealth investors over the last decade. However, almost all have gone into existing planted and managed native forests, and have not led to expanding the area of planted forests, with a few, very minor exceptions in Latin America.

Historically, the expansion of planted forests correlates directly with consistent government support coupled with affordable land purchase/rent costs and identified markets for wood products. The alignment of these positive factors in several countries which occurred from 1960 to 2000 is no longer so obvious. The great expansions in both government and private-sector forest areas in South Africa, Australia (softwoods), New Zealand, Chile, the USA, the UK and elsewhere have diminished or effectively stalled.

The Chinese government is clear in its intention to expand planted forests. It is notable, however, that the area of 'timber planted forests' in China actually reduced by 4.1% between 1998 and 2003. During the same period the area of 'other' (probably protection) planted forests, such as the Three North Shelterbelt Programme, increased by 82%.[5] In 2007 RISI predicted that China's 'operable timber planted forests' will increase from about 25.5 million hectares in 2005 to about 29.5 million hectares in 2020. The generous tax breaks for planted forests in Australia are encouraging expansion, though already land prices may be limiting affordability.

Future demand for pulpwood fibre is driving expansions of planted forests in Brazil, Laos, Malaysia and Vietnam. Continuing expansion of planted forests in Indonesia is on hold while the Department of Forestry and the Police reassess the legality of all existing permits to convert native forest land to planted forests.[6] Such land conversion is controversial, and has been widely criticized by environmental groups and NGOs.

Several companies and investors are considering major planted forests programmes in sub-Saharan Africa. At least 25 million hectares of land is suitable for planting south of the Congo River[7] where annual rainfall exceeds 1000 mm on slopes too steep for large-scale food production. Such afforestation would add massively to Africa's planted forests, but political policies, corruption and other issues have to date severely limited turning this potential into viable projects.

For decades planted forest investment, whether undertaken by governments or private enterprise, was driven by 'patient' funds. Investors accepted periods of 20, 30 or even 50 years between planting and harvesting. This is no longer the case. Most investors now seek to harvest trees within 5–15 years. Future planted forests will be for shorter rotation crops, and for processing technologies to add value to fibre and solid-wood from young trees.

Continuing investment in existing planted forests by institutional funds can be expected, but it will focus on re-arranging ownership of existing assets rather than augmenting significantly the global planted forest resource. Some loss of planted forests to HBU seems likely.

Increasing competition from major agri-business investing in palm oil, corn and maize, and from rapidly expanding biofuel crops such as *Jatropha*, are likely to drive up global land prices and rents. This trend may limit planted forest expansion, a phenomenon already happening in Indonesia, Malaysia and Brazil, while in Thailand farmers are being paid more for growing tapioca than short-rotation *Eucalyptus* tree crops.

The continuing loss of native forests in Africa, Asia and Latin America, in spite of the best efforts to slow the rate of destruction, may lead to increases in wood prices from planted forests. It is already apparent that available native forest resources in Indonesia and Malaysia have diminished rapidly since 2000, and the large-scale harvesting of native forests in Africa to feed Western, and more recently, Chinese industries will have the same effect within 5–10 years. Changes to tax policies in major wood-producing countries, e.g. Russia, may also lift wood prices. Entrepreneurial efforts to raise funds for multi-use planted forests, including wood, carbon and biofuel values, are likely to increase establishment rates somewhat, even if obtaining suitable land becomes more difficult. The rapidly growing Indian economy will require more wood resources in future. To date restrictive government policies have limited the expansion of planted forests, but this may change. In the meantime, some Indian companies are expected to expand planted forest areas in countries such as Malaysia and Laos.

Finally, regional political blocs may become a new catalyst for planted forest expansion as part of goals and targets for reducing greenhouse gas emissions. For example, the Asia Pacific Economic Cooperation Forum (APEC), in their agreed target of reducing 'energy intensity' by 25% by 2030, pledged to increase forest cover in the region by at least 20 million hectares.

7.3.12 Conclusions concerning the corporate private sector's investments

- Over the past 50 years, most countries that have expanded their sustainable, commercial planted forest estate have had key characteristics in common: stable governments, strong security of land tenure and an independent judiciary to protect investor rights. Almost all provided generous tax relief or subsidies over long periods of time. These may have been discontinued, or modified once a target area had been established.
- The most effective way international institutions such as the FAO and the World Bank can ensure similar expansion of private planted forest investments in developing countries is to assist governments to improve security of land tenure, and to ensure that investors are protected by independent legal systems.

- The ownership of planted forests in many countries in North America, Oceania and Latin America has been transformed in the last 30 years, and particularly in the last 5–10 years. Governments and integrated forest product ownership no longer dominate; institutional and private equity funds are now big players. Additional investors and funding sources are entering the sector.
- There are many initiatives from governments and political blocs to increase substantially the area of planted forests, largely driven by the global warming and carbon sequestration issue. However, the mechanisms to secure the necessary land and specific funding sources remain unclear.
- Planted forest expansion will face increasing competition, in both developed and developing countries, from food production and the emerging biofuels industry.

7.4 Stakeholder Issues

Trees and forests by their very nature – their scale and influences – attract interest and concern far beyond that of their owners. Many stakeholders, from those deprived of light or moisture by trees on adjacent land to those critical of land carpeted by monocultures, feel threatened. Many stakeholders gain direct employment or possess rights to exercise. Many stakeholders are investors or derive income from planted forests. And the public at large see their landscape, environment, even their water supply affected by planted forests to name but a few non-market impacts. All these interests, and many others, are now recognized if not all accorded equal or at least appropriate engagement. Ensuring engagement with stakeholders in ways that are just and sound is at the heart of responsible management of planted forests and are spelt out in FAO's voluntary guidelines (FAO, 2006b). Compliance with them is reviewed with care in the certification process – see also poverty issues in the previous chapter.

For planted forests, beyond those in smallholder ownership, structured engagement with stakeholders is imperative. Higman *et al.* (2005) indicate why great emphasis is now needed, why social issues are important, and the steps to take to work with the many different stakeholder groups. Evans and Turnbull (2004) illustrate successful examples for tropical planted forests, such as in the many community forestry projects in Ethiopia (Fig. 7.3). Farmers whose land was planted, village leaders, NGOs in the form of a local church and an international Christian charity, and local government all participated to a greater or lesser extent as appropriate in deciding land allocation, annual work programmes, day-to-day issues of food-for-work provision, shepherding to control livestock grazing, and (later) harvesting and use of trees for fodder, fuel and building poles. In the UK all forests owners' plans for tree felling are open to public scrutiny and comment by interested bodies concerned with wildlife conservation, archaeology, planning issues etc., as well as the requirement to comply with laws and guidelines aimed at protecting the public good.

Fig. 7.3. Women filling plastic tubes to grow seedlings as part of a community forestry project to grow fuelwood and prevent soil erosion, Gamo Gofa, Ethiopia, which involved the villagers, local government officials, a local church and an international NGO.

7.4.1 Issues specific to planted forests

Several stakeholder issues arise from the trend towards privatization. Private or corporate ownership of extensive resources may be less sympathetic towards local concerns and lead to displacement of people (as happened with some communities in South Africa (Dlomo and Pitcher, 2005)), to changed power structures in local communities, even to the capacity of local people to be involved. Such failure to engage with stakeholders and not responding to genuine concerns can lead to the planted forests themselves being targets for grievances through arson or deliberate encroachment of forest land. This is particularly the case in the less regulated and organized societies of many tropical countries (Evans, 2001a; Evans and Turnbull, 2004; FAO, 2006b).

Some issues are more a matter of perception. For example, corporately owned planted forests threaten local livelihoods rather than bring employment opportunities; they damage environmental quality rather than protect soil, sequester carbon, or protect enclaves or corridors of native forest; or they benefit from subsidies and tax breaks rather than contribute to GDP (including through taxation) and generate wealth. Such perceptions may flow from dislike of corporate dominance and, perhaps, globalization generally, but they are no less genuinely felt. As with so many issues, balance is needed. But as Garforth *et al.* conclude about this trend 'Plantation transfer is not about trees; it is about people and power' (2005b, p. 98).

Conflicts among agencies and departments – the stakeholders representing government and national interests – are a real issue. Lack of coordination can lead to failed compliance in one aspect as owners or managers strive to satisfy another. The nature of planted forests as an extensive land use impacting livelihoods and the environment in countless ways inevitably encounters a raft of measures and regulations.

7.5 Regulation, Certification and Related Issues

The simpler type of forest conditions and manageable scale that planted forest possesses have rendered them amenable to certification. The vast majority of currently certified forest areas (both tropical and non-tropical) are industrial forests (Durst *et al.*, 2006), most of which will fall within the definition of planted forests. In countries such as South Africa and the UK, the great bulk of the forest estate that is certified is planted. The process is seen as conditional on acceptance by the public that forests are managed sustainably (Higman *et al.*, 2005) and in the UK is described as a 'runaway success' because of the consensus achieved among the key stakeholders (Wilson, 2007). That independent certification will expand further and affect virtually all planted forests appears a given at the present time: it is becoming the principal tool for ensuring high standards of development, management and operational practice.

The exception to the above may be where there is no culture of forest stewardship, as is argued in the case of China (Stone, 2006), because certification takes conscious and dedicated management. Where ownership and usage rights are divorced or where, as noted earlier, investors and managers of a particular planted forest resource are driven by different objectives – one perhaps by a purely financial return, the other a greater focus on sustainable production of timber over time – ensuring compliance will be difficult without stringent enforcement practices.

The relation to trade is less clear. Forest products from non-certified sources are very gradually becoming less marketable, particularly in Western economies, but still the proportion of the potential annual supply of certified wood actually traded as certified products is small (Durst *et al.*, 2006) and there is limited evidence of a price premium. It is doubtful whether better prices will cover the costs of certification, though improved sales (better market access) may result, thus the trade dimension is less of a driver of certification.

Because compliance increasingly forms part of the regulatory framework, certification seems likely to evolve into setting minimum standards rather than a mechanism delivering mainly trade advantage. While it undoubtedly helps the environment, certified timber is more like food that meets a state's minimum criteria for safe human consumption than organic produce attracting a premium price as being better for you as well as for the environment. With certification perceived as part of the regulatory environment, the question arises of governments using it as a proxy for their own controls (Garforth *et al.*, 2005b), a development specifically encouraged in South Africa though with local adaptations (Scotcher,

2006). In the last 15 years certification has had a huge impact, but as a tool to drive up standards it is still in the development phase.

7.5.1 Some certification issues

The certification process is costly and, particularly for small owners, is a burden per hectare (overhead) on the already meagre returns most forestry offers. While group certification remains an option (Nussbaum, 2002; Higman *et al.*, 2005), the greater burden is a form of discrimination (Mayers and Bass, 2004) and market distortion. The cost arises from the necessary bureaucracy, even when kept to an absolute minimum, and the obvious requirement of monitoring compliance through regular reports and inspections. As the process becomes established, acceptable economies will certainly arise, but the additional overhead seems unavoidable.

Compliance may bring greater restriction in the future if certain practices are deemed unacceptable, such as use of genetically modified planting material or a ban on pesticides. These constraints largely derive from NGO-led public pressure on intensive agriculture. This is unfortunate for planted forests since wood and timber products are not for human consumption and hence of no threat to health, and pesticide use is minute, both in quantity and frequency, compared with farm applications. Annual pesticide application on all forests in the UK only equals that which is applied to the gardens of one medium sized town! Nevertheless, the forest sector must recognize that their practices impact the wider environment, and none would disagree with the desirability to minimize pesticide use and to acknowledge the potential threat that all artificial genes represent to natural ecosystems.

7.5.2 Certification schemes

Several certification schemes have arisen, and while different groups advocate particular ones as most closely meeting their aims, the differences between them will lead to greater or lesser emphasis on certain issues. Some are more market orientated, others towards environmental and social imperatives, and still others compliance with laws and regulations. While some diversity can be welcomed, a likely trend is to develop national forest management standards that reconcile or are compatible with more than one certifying body, such as Greece's approach to FSC (Forest Stewardship Council) and PEFC (Programme for the Endorsement of Forest Certification) (Georgiadis and Cooper, 2007). The inability to achieve this in many developing countries is a major obstacle to increasing forest certification where it is most needed (Durst *et al.*, 2006).

As Bass concludes: 'Forest certification is based on concerns of both global and local imperatives for sustainable forest management and reflects the ongoing process of negotiation of the often conflicting ideas of what sustainable forest management is about. Where certification can manage these tensions creatively, it should certainly have an enduring role' (2004, p. 1356). This is nowhere more true than for planted forests and the arguments surrounding their continuing and increasing role.

7.6 Compliance, Illegal and Illicit Practices

In general, compared with native forest, planted forests suffer less from illegal practices. Ownership, investment and economic activity, and higher levels of management, all bring a greater degree of care and less opportunity for some types of crime, particularly that facilitated by remoteness and distance from supervision. Both the fact of certification and the generally easier monitoring of compliance with codes of practice, regulations and laws suggest that the whole subject of illegality will have a less high profile in the outlook for planted forests compared with native ones. It is more likely that the compliance challenges for the future will be ensuring that social, cultural, environmental and economic principles, guidelines and practices (e.g. FAO's *Responsible Management of Planted Forests*; FAO, 2006b) become realities in policies, planning and, crucially, field implementation.

Notes

1 www.risiinfo.com
2 *The Global Tree Farm and Managed Forest Industry Review – 2007 edition.*
3 'Timberland' is the American term for 'forestland'.
4 Bioenergy International.
5 China's National Forest Inventory.
6 Industry sources.
7 SAPPI, 2002.

8 Sustainable Silviculture and Management

J. Evans

8.1 Introduction

This penultimate chapter examines issues that may enhance or threaten the outlook for planted forests. It reviews silviculture and management in planted forests, focusing on biological and environmental aspects. Are there threats or opportunities that can be anticipated and so guide the direction of planted forests as they increasingly dominate supply of forest products as well as play other roles? Because the act of planting forest deliberately brings greater uniformity, are there attendant risks as well as the advantages of simpler management? On the one hand, are planted forests 'sustainable' or are there flaws biologically or even artificially, such as fires arising from arson, which will eventually lead to insuperable problems for such silviculture? On the other hand, are there new directions that may enhance provision of forest products and services from the world's planted forests?

The converse must also be addressed. Do planted forests themselves pose a threat, for example, to soil fertility, or from the risk of exotic species becoming invasive, or over-zealous development and management damaging other ecosystems, as is happening to some tropical rainforests cleared for biofuel crops? Some of these issues were touched on in Chapter 6. Here, mainly the biological and ecological processes of planted forests are considered.

The chapter follows a logical sequence of questions to ask of planted forests in the context of their future and outlook. It updates and broadens the analyses in Evans (1999, 2001b, 2004a).

- Are planted forests sustainable?
- Do they induce changes to a site, and are such changes harmful or beneficial?
- Is there any evidence of problems if the same site is re-used again and again as rotation succeeds rotation?
- What risks are planted forests exposed to, in particular are they stable ecologically?

- How can production and other benefits be enhanced?
- What threats do planted forests themselves pose?
- What about trees on farms and other non-forest plantings?

8.2 Are Planted Forests Sustainable?

The question of sustainability has two components. There are the broad issues of the two previous chapters of whether using land and devoting resources to planted forest is a sustainable activity from the economic, environmental or social dimensions. Is such development ultimately unsustainable and a threat rather than a help to people's livelihoods and land use? These, and related issues, are important and fundamentally depend on national policies governing and regulating planted forests and their development, understanding their impacts, and ensuring full stakeholder participation in the process.

The second component, sometimes labelled 'narrow-sense' sustainability, is largely a silvicultural and management issue. The question raised is: can planted forests be grown indefinitely for rotation after rotation on the same site without serious risk to their well-being? More specifically, can their long-term productivity be assured, or will it eventually decline over time? These questions are pertinent owing to the increasing reliance on planted forests, but are also scientifically challenging, since in previous centuries trees and woodlands were seen as 'soil improvers' and not 'impoverishers'. Are today's silvicultural and harvesting practices more damaging because of greater intensity and the high timber yields achieved compared with that from most native forest? And, of course, are resources such as genetic improvement, targeted fertilizer application and sophisticated manipulation of stand density, along with rising atmospheric carbon dioxide, likely to lead to crop yield improvement, or could they disguise evidence of genuine site degradation or increasing risk of damaging pests and diseases?

Understanding sustainability also applies to non-industrial uses of planted forests, including those largely devoted to protection purposes. Sustaining the numerous benefits people derive from planted trees should be a top priority and arise out of good management. Does, for example, the perpetual gathering and removal of leaves, twigs and litter from beneath tree stands, so widespread in India and China, simply loot the site of nutrients? And what of the flow of non-timber products, often of more value than wood, and perhaps less directly damaging to sites when harvested? These are relevant to planted forests, even if it is not always possible to answer such questions adequately.

The next two sections look at evidence worldwide to address two elements of narrow-sense sustainability.

1. What changes to a site may planted forests induce that may threaten future rotations?
2. What factual evidence is there for and against productivity change over time?

8.3 Do Planted Forests Induce Site Change?

Two important questions are: (i) do the silvicultural practices commonly applied, such as planting exotic species, use of monocultures, clear felling systems etc., cause site change; and (ii) are such changes more or less favourable to the next crop?

This is a much researched topic and summaries of the science were presented in the 1990s by Dyck *et al.* (1994) in *Impacts of Forest Harvesting on Long-Term Site Productivity*, Nambiar and Brown (1997) in *Management of Soil, Nutrients and Water in Tropical Plantation Forests*, Evans (1999) *Sustainability of forest plantations – the evidence* and, more recently, for planted forests in Brazil by Gonçalves *et al.* (2004a) in *Forest Nutrition and Fertilization*.

8.3.1 Assessing changes in soil

Demonstrating that soil changes because of forestry practices is difficult to establish conclusively both in fact and in scale (Powers, 2001). An absence of sound baseline data is common; moreover, is the reported change actually induced by the planted forest?

The second question is whether the observed changes represent degradation or improvement. There are remarkably few examples of changes supposedly induced by *growing* trees that led to less favourable conditions for that species, and they are usually unintended (Burger, 2004). Equally, the irreversibility of changes has rarely been demonstrated, apart from obvious physical losses, such as erosion of topsoil. A gradual trend, perhaps observed over several decades, can be quickly reversed as stand conditions change. As Nambiar pointed out, 'the most striking impacts on soils and hence productivity of successive crops occur in response to harvesting operations, site preparation, and early silviculture from planting to canopy closure (1996, p.1631).'

Inadvertent bias, however, can come to dominate. Investigation of changes is often carried out specifically because there is a problem which has already revealed itself in poor tree growth or health. Also, forest soils are notoriously variable, owing to the kind of land commonly used for planting, making single plots or small sample comparisons often wildly inaccurate. Little recognized, too, is that measured values of many soil parameters change radically from season to season. These issues need addressing in impact studies of planted forest on soils (Evans, 1999). Drawing conclusions from limited investigations covering only a few years of a rotation can grossly mislead if extrapolated over whole rotations and beyond.

8.3.2 Soil chemical status

Planted forests may remove nutrients from soil as trees grow and then are harvested, and may change soil surface chemistry as the litter layer and organic matter are dominated by one species with uniform composition and decay characteristics. Furthermore, site preparation practices, such as ploughing, drainage and fertilizing, directly affect soil physical parameters and in turn nutrient and moisture availability.

Soil as a mineral store

Soils vary enormously in their role as a nutrient reservoir. Whereas arable farming treats soil as a medium for growing crops with nutrients largely supplied by fertilizers, in forestry the perennial and generally deep rooting of trees focuses less on soil reserves and more on where the dynamics of nutrient supply is mediated – i.e. the soil surface. Indeed, forests, including planted forests, are highly efficient recyclers of nutrients and almost 'leak free' if undisturbed. In the tropics, where recycling can be at its most efficient, nutrients in mineral soil often no longer represent the dominant proportion of the ecosystem (Evans and Turnbull, 2004). The soil – in many temperate as well as tropical regions – often plays only a small part in nutrient exchange and it is the surface organic, root-bearing zone, especially the annual turnover of fine roots, which is important in concentrating energy flow from decomposing organic matter back into living organic matter (Briggs, 2004). The integrity of this layer and how it is handled in planted forests is critical. As Attiwill and Weston (2001) emphasize, availability of nutrients in forest soils is a long-term concern that should address biological processes, as well as inorganic equilibria as in farm soils, over years and even centuries.

Nutrient removal

Nutrient removal from planted forests occurs when any product is gathered or harvested, such as leaves, fruits, litter, logs or whole trees. Many studies have been made: Gonçalves *et al.* (1997) alone list 12 tropical examples and Morris (2004b) concisely summarizes nutrient cycling. Critical to forest sustainability is what proportion the nutrients lost represent of the whole store. This ratio of nutrient export:nutrient store is advocated as a measure of long-term ecosystem stability (though the questions remain, what is the store and how can it be measured?). Typical rates of removal are often 10% or less of the soil store, certainly in most temperate ecosystems, i.e. a stability ratio of <0.1 (Attiwill and Weston, 2001). An example of an exception is Folster and Khanna's (1997) data for *Eucalyptus urophylla* × *grandis* hybrid stands with three very different site histories at Jari in North-Eastern Amazonia, Brazil. To quote: 'Twelve of the stands were in the second to fourth rotation, indicating that most of the previously grown *Gmelina*, *Pinus* or *Eucalyptus* had already extracted their share of base cations from the soil and left it greatly impoverished' (Folster and Khanna, 1997, p. 350). The stability ratio was >1, suggesting unstable and therefore unsustainable conditions. Caution, however, is needed since others (e.g. Rennie, 1955; Binns, 1962; Johnson and Todd, 1990) have long predicted calcium shortages from comparison of removals in harvested biomass with available quantities in the soil, yet trees continue to grow where conventional soil analysis suggests there is virtually no calcium!

The size of nutrient losses depends, crucially, on what parts of the tree are removed – debarked log, log with bark, whole tree including branches etc. – owing to the highly unequal distribution of nutrients in plant tissue. In general, if the stability ratio (above) is greater than 0.3, there may be serious sustainability questions in the longer term, and if it is above 0.5 in the immediate future. Understanding these dynamics helps identify at what points on the continuum of growth across planted forest sites, species and productivities the ratio becomes critical for long-term stability. There appear to be few examples of reaching such limits. It is worth remembering that nutrient removals by forest crops are typically only one-fifth to one-tenth that of arable farming (see Miller, 1995), and that both re-translocation of nutrients within a tree – from dying organs to new growth – and their recycling via the litter and soil surface are the most significant dynamics taking place (Miller, 2004).

Litter and residues

The influence of litter on soil chemical status may be important, since needles and leaves of different species vary in composition and decay at different rates. For example, in Southern Africa substantial accumulations may develop under *Pinus patula* on certain sites (Morris, 1993), while this is rare beneath the more lightly canopied *P. elliottii*. In broadleaved stands accumulation of litter is uncommon, though not unknown, e.g. under some *Fagus* and *Quercus* stands on acid soils in Europe. Even under *Tectona grandis* and *Gmelina arborea*, which usually suppress all other vegetation, the large leaves readily decay. Similarly, build up is rare under the light crowns of *Eucalyptus* and *Fraxinus* and the nitrogen-rich foliage of leguminous trees such as *Acacia*, *Leucaena* and *Prosopis* spp. and non-legume nitrogen-fixers such as *Alnus* and *Casuarina*, owing to rapid decay of the rich organic matter.

Measured changes in soil chemistry

The above processes indicate that planted forests could influence soil chemical status. Most studies have either compared planted and non-planted sites or examined trends as a planted stand develops. Few have examined changes over successive rotations and even fewer comparisons between planted forests and farmland.

In both temperate and tropical studies, *increases* and *decreases* in carbon, nitrogen and macronutrients under stands in planted forests compared with native forests or pre-existing conditions have been reported – see references in Evans (1999) and for the hugely important carbon stocks, Dawson and Smith (2007). Not surprisingly, nitrogen accumulation is widely found under nitrogen-fixing species.

Numerous studies in temperate forests have focused on pH change, litter type, podzolization etc. and the impacts of acid rain, though distinguishing the latter from direct tree effects on soil acidity is difficult. On the whole, tree impacts are relatively small compared with the soil nutrient store, the underlying geology and deposition of aerial pollutants.

8.3.3 Soil physical condition

Planted forests may affect soil physical conditions through: (i) site preparation and establishment operations; (ii) the effects of tree growth itself; and (iii) harvesting practices.

Site preparation and planting

Cultivation and drainage affect soil physics for many years and sometimes for more than one rotation. Obviously site preparation seeks to improve growing conditions for trees and not impair sustainability or productivity. Longer term benefits include reduction in bulk density, increased infiltration capacity and soil aeration, improvement in moisture storage and enhanced mineralization rates of organic matter. Physical disruption of indurated layers and deep cultivation such as tining are actually designed to reverse 'undesirable' soil profile development.

Impact of tree growth

The general conclusion about water use by trees was noted in Chapter 6, namely, that compared with many land uses, trees exhibit higher evapotranspiration, i.e. they 'use' more water. On some sites this is harnessed to dry them and lower the water table: there are instances of *Eucalyptus* planted for this purpose (Evans and Turnbull, 2004). However, it is difficult to quantify this effect on the growth of later rotations. If a planted forest loses more moisture than is received by the site in precipitation, no soil moisture recharge will occur and reserves are depleted. In the US mid-West, some forests planted in the early 1900s thrived initially, but died once moisture reserves were used up and precipitation was inadequate to sustain growth (Kramer and Kozlowski, 1979).

The converse is also commonly observed. When a stand of trees is felled, re-wetting of the surface may be evident (Savill *et al.*, 1997). The effect is usually transitory and disappears at or before canopy closure of the next crop.

Indirect impact of vegetation suppression

Planted stands of *Tectona* and *Gmelina* in the tropics, and many conifers in both tropical and temperate conditions, may suppress ground vegetation. When this exposes the soil, perhaps because litter is burnt or gathered, erosion rates increase. The protective function of tree cover derives more from the layer of litter and organic matter at the soil surface than from interception by the canopy. In India, raindrop erosion was nine times higher under planted *Shorea robusta* stands if the litter had been lost through burning (Ghosh, 1978). Soil erosion beneath planted *Paraserianthes falcataria* was only 0.8 t ha/year where litter and undergrowth were intact, but an astonishing 79.8 t ha/year where it had been removed (Ambar, 1986). Wiersum (1983) found virtually no soil erosion under planted *Acacia auriculiformis* when litter and undergrowth were intact, but serious erosion where local people gathered the litter.

Harvesting damage

Extracting logs can cause soil compaction, scouring of the soil surface and erosion, blocking of ditches and other drainage channels, and oil spillage. The method of extraction greatly influences the extent of damage, with draft systems using mules, oxen etc. being least harmful, and skidding with wheeled and tracked vehicles generally most damaging. Both the weather and type of soil affect the severity of damage, with compaction worst when extraction occurs in wet conditions on heavy clay soils.

There are many reports of impaired growth of trees planted on extraction routes and where soil has been compacted or suffered erosion: a useful summary is in Nambiar (1996).

8.3.4 Organic matter dynamics

As noted earlier, the litter and organic matter layer may influence soil surface chemistry, but this role of the forest floor is even more critical to sustainability of planted forests because:

- it helps prevent soil erosion
- it improves infiltration of moisture
- it represents a significant nutrient store, albeit a dynamic one (Morris, 2004a) and
- the litter–humus–mineral soil interface is the seat of nutrient cycling and microbial activity.

Consequently, minimizing disturbance of the forest floor, especially during harvesting operations, is an important goal of forest management (Briggs, 2004; Moffat *et al.*, 2006). Any activity that disturbs these roles of organic matter in the ecosystem can have large effects, and perhaps the most serious activity, still practised in some countries, is regular litter raking or gathering (Fig. 8.1). For planted forests the cost of managing debris and site preparation when restocking is expensive and a high proportion of establishment costs, but as Nambiar pointed out, 'one shoddy operation can leave behind lasting problems' (1996, p.1641).

Fig. 8.1. Litter raking under *Pinus caribaea* in southern China.

Where yield decline over successive rotations has occurred (see later), part of the cause usually includes harmful practices regarding litter and organic matter. And if new wood fuel strategies (such as in the UK aimed at mitigating climate change by saving fossil fuels) renew scavenging of forest residues and removal of branches and litter, concern over sustainability may re-awaken. In Brazil minimum soil disturbance regimes are now commonplace in planted forest silviculture to conserve organic matter, to reduce nutrient losses and soil erosion, and elevate biological activity, which incidentally reduce establishment costs and weed problems (Gonçalves *et al.*, 2004b).

8.3.5 Weed spectrum and intensity

Establishment of forests by planting greatly affects ground vegetation, with some operations designed to reduce weed competition. Weed control ensures that the planted trees have sufficient access to site resources for adequate growth. Once canopy closure has occurred, weed suppression is usually achieved for the rest of the rotation. A critical next phase is managing weeds through the harvesting and restocking process in re-establishing the crop.

In subsequent rotations the weed spectrum often changes. Owing to past weed suppression, exposure of mineral soil in harvesting, and the accumulation of organic matter, conditions for weed species change. Birds and animals may introduce or spread new weed species, grass seed may be blown into stands and accumulate over several years only to flourish when the canopy is removed. Roads and tracks in planted forests can become sources of weed seeds. Weed management must be a holistic operation. As with a failure to handle organic matter carefully, where yield declines have been reported, often the significance of weeds has been insufficiently recognized on restocked sites in second, third and later rotations (Powers, 2001; Evans, 2004b).

8.4 Evidence of Productivity Change

8.4.1 Data from successive rotations of planted forests

For planted forests, there is meagre hard evidence of productivity change over successive rotations, with few reliable data. The long cycles in forestry make data collection difficult. Records are rarely maintained from one rotation to the next; funding for long-term monitoring is often a low priority; detection of small changes is difficult; and often the exact location of sample plots is poorly recorded (Evans, 1984). Moreover, while many older planted forests have second rotation stands, few have third or later rotations, and so the opportunity to collect a run of data has been limited.

The few comparisons of productivity between rotations have mostly arisen from concern over yields, the fabled 'second rotation decline', or about stand health. The focus has been on problems: across the vast extent of planted forests none has arisen, suggesting no great concern, and managers are not encountering obvious decline problems. Thus data available in the older literature may be biased to problem areas, while more recent studies may be less so, such as the European Forestry Institute survey (Spiecker *et al.*, 1996) and CIFOR's project 'Site management and productivity in tropical forest plantations', which incorporates systematic establishment of sample plots. Nevertheless, any analysis of the outlook for planted forests must consider the cases of decline that have arisen.

8.4.2 Review of evidence comparing yields in successive rotations

Six major studies have investigated productivity in successive rotations, and there is also some anecdotal evidence and occasional one-off studies. These are grouped by region.

Picea abies *in Saxony and other European evidence*

As was noted in Chapter 2, in the 1920s Weidemann (1923) reported that significant areas of second- and third-rotation *Picea abies* in lower Saxony (Germany) were growing poorly and showed symptoms of ill-health. In 8% of stands there was a fall of two quality classes in second- and third-rotation stands. It is now clear that this mainly arose from planting spruce on sites to which it was ill-suited. Today, young stands of the species in Saxony and Thuringia are growing more vigorously than their equivalents 50 or 100 years ago (Wenk and Vogel, 1996).

Elsewhere in Europe comparisons between first and second rotations are limited. In Denmark Holmsgaard *et al.* (1961) reported no great change for either *P. abies* or *Fagus sylvatica*, though more recently second-rotation *F. sylvatica* is growing significantly better (Skovsgaard and Henriksen, 1996). In The Netherlands, second-rotation planted forest generally grows 30% faster than the first (van Goor, 1985) and in Sweden second-rotation *P. abies* shows superior growth (Elfling and Nystrom, 1996). In France decline has been reported in successive rotations of *Pinus pinaster* in the Landes, though it is not attributed to site deterioration (Bonneau *et al.*, 1968). In the UK most second-rotation stands are equal to or better than their predecessor, and no decrease in growth expected (Savill *et al.*, 1997).

Pinus radiata *in Australia and New Zealand*

Significant yield decline in second-rotation *P. radiata* occurred in South Australia in the early 1960s (Keeves, 1966), with a worrying 30% drop. Near Nelson in New Zealand, transitory second-rotation yield decline occurred on a few impoverished ridge sites (Whyte, 1973). These reports were alarming and generated a great deal of research. By 1990 it was clear for South Australia that harvesting and site preparation that failed to conserve organic matter and a massive influx of grasses in the second rotation were the main culprits. By rectifying these problems and using genetically superior stock, second- and third-rotation *P. radiata* now grow substantially better than the first crop (Boardman, 1988; Woods, 1990; Nambiar, 1996).

Elsewhere in Australia the second rotation is equal or superior to the first – see a summary in Evans (1999). In New Zealand, on the great majority of sites, successive rotations gain in productivity, though Dyck and Skinner (1988) conclude that inherently low-quality sites that are managed intensively will continue to be at risk.

Pinus patula in Swaziland

Long-term productivity research in the Usutu forest, Swaziland, began in 1968 as a consequence of second-rotation decline reports from South Australia. For 40 years measurements have been made over four successive rotations of *P. patula* from a forest-wide network of long-term productivity plots.

The most recent analysis in Evans (2005) shows second, third and fourth rotation height-growth at age 6 years from plots growing on exactly the same sites. (First-rotation growth data were derived from stem analysis and from paired plots, and are less accurate). Table 8.1 compares the second and fourth rotation, where there are sufficient plots for a statistical comparison. Table 8.2 provides data based on a smaller number of plots, but where it is possible to compare all three rotations. A full analysis of the first three rotations up to rotation age (15–17 years) will be found in Evans and Masson (2001).

Table 8.1. Comparison of mean height of *Pinus patula* at age 6 years in second and fourth rotations on exactly the same sites (based on 41 permanent sample plots).

	Unadjusted data		Adjusted data	
Rotation	Height (m)	Variance	Height (m)	Variance
2R	7.48	1.24	7.87	1.12
4R	8.15	1.16	8.64	0.89
't' statistic	3.20		4.52	
Significance	P>0.001		P>0.001	
% change	+9.0		+9.8	

Adjusted data incorporate minor modifications to reflect differences between rotations in incidence of hail damage, month of planting and similar complicating factors.
Source: modified from Evans (2005).

Table 8.2. Comparison of mean height of *Pinus patula* at age 6 years in second, third and fourth rotations on exactly the same sites (based on 24 permanent sample plots).

	Unadjusted data		Adjusted data	
Rotation	Height (m)	Variance	Height (m)	Variance
2R	8.03	1.15	8.34	1.05
3R	8.14	1.12	8.23	1.19
4R	8.61	1.00	9.12	0.71
% change 4R:2R	+7.2		+9.4	
% change 4R:3R	+5.8		+10.8	

Adjusted data incorporate minor modifications to reflect differences between rotations in incidence of hail damage, month of planting and similar complicating factors.
Source: modified from Evans (2005).

These tables and the data in Evans and Masson (2001) show that each successive rotation is equal to or a little better than its predecessor (Fig. 8.2a). On a small part of the forest (about 13% of its area), on phosphate-poor soils, a decline did occur between first and second rotation, but this has not continued into the third rotation (Fig. 8.2b).

The importance of the Swaziland data, apart from the long run of measurements, is that no fertilizer or other ameliorative treatments have been applied to any long-term productivity plot from one rotation to the next. Some third rotation *P. patula* may have benefited from genetic improvement (Morris, 1987) as will have some plantings of the fourth (Evans, 2005). However, Swaziland has suffered severe droughts along with the rest of southern Africa (Hulme, 1996) and hail storms that will have adversely impacted more recent growth (Evans,

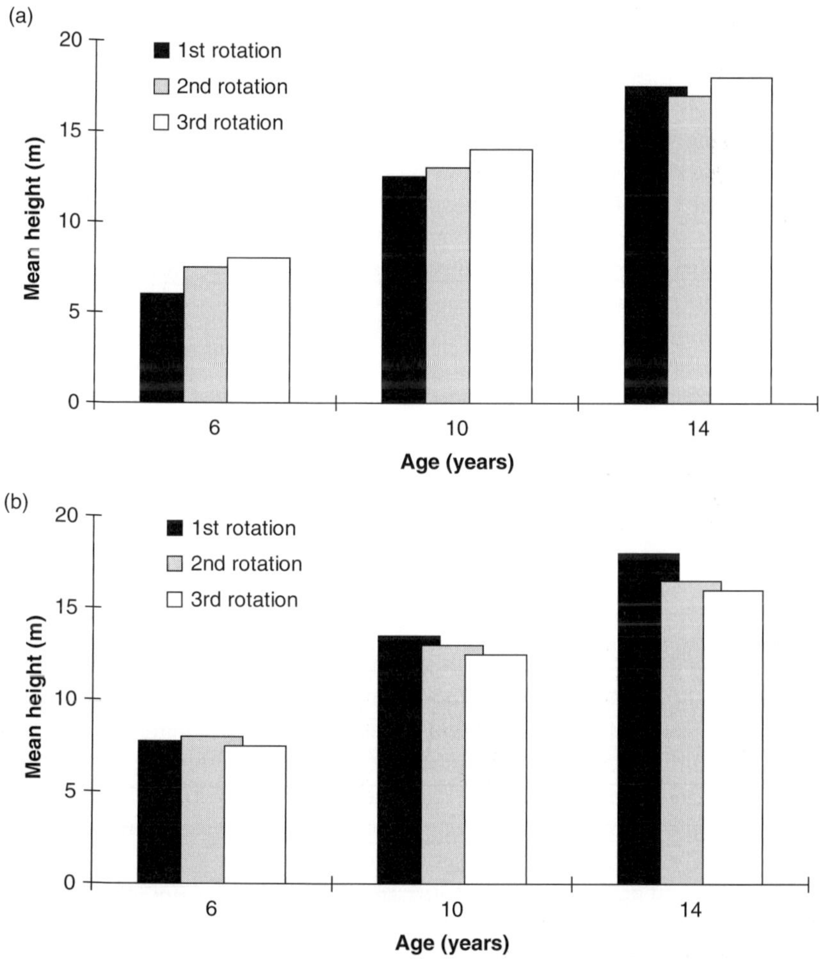

Fig. 8.2. Mean height growth (m) of *Pinus patula* over three rotations in the Usutu Forest, Swaziland. (a) most of the forest planted on granite derived soils; (b) 13% of the forest on phosphate-poor gneiss soils.

2005). Assessments are continuing to see how the later growth in the fourth rotation is affected.

These data are of interest because the silviculture practised over a planted forest of 62,000 ha is intensive: monoculture, no thinning or fertilizing (except to correct phosphate deficiency over a small part of the forest; Morris, 2003), and a rotation of 15–17 years which is close to the age of maximum mean annual increment. Large coupes are clearfelled and all timber extracted for pulpwood. Slash is left scattered (i.e. organic matter conserved) and replanting done through it at the start of the next wet season. The limited genetic improvement of some third and fourth rotation stands could have disguised a small decline, but evidence is weak. Overall, the data suggest no serious threat so far to narrow-sense sustainability.

Cunninghamia lanceolata *in sub-tropical China*

There are about 6 million hectares of planted *C. lanceolata* in subtropical China. Most are monocultures and worked on short rotations to produce small poles, though foliage, bark and even roots are harvested for local use. Reports of significant yield decline have a long history. Ding and Chen (1995) and Ying and Ying (1997) report a drop in productivity between first and second rotation of about 10–30%, and between second and third rotation of up to 40%. Chinese forest scientists attach much importance to the problem and pursue research into monoculture, allelopathy and detailed study of soil changes. However, the widespread practices of whole tree harvesting, total removal of all organic matter from a site and intensive soil cultivation that favours bamboo and grass invasion all contribute substantially to the problem (Fig. 8.3). Ding and Chen conclude that the problem was 'not Chinese fir itself, but nutrient losses and soil erosion after burning (of felling debris and slash) were primary factors responsible for the soil deterioration and yield decline . . . compensation of basic elements and application of phosphate fertilizer should be important for maintaining soil fertility, and the most important thing was to avoid slash burning . . . . These (practices) . . . would even raise forest productivity of Chinese fir' (1995, p. 66).

Tectona grandis *in India and Indonesia*

In the 1930s evidence emerged that replanted teak (*T. grandis*) forest was not growing well in India and Java (Griffith and Gupta, 1948). Although soil erosion is widespread under *T. grandis* and loss of organic matter through burning leaves is commonplace, research into the 'pure teak problem' did not generally confirm a second-rotation problem. However, Chacko (1995) and Chundamannii (1998) describe site deterioration as still occurring, with yields from planted stands not up to expectation and a general decline of site quality with age. Causes are attributed to poor supervision of planting and establishment, over-intensive taungya (intercropping) cultivation, delay in planting, and poor maintenance and management generally.

In Java, Indonesia, where there are about 600,000 ha of *T. grandis*, site deterioration is a problem and 'is caused by repeated planting of teak on the same sites' (Perum Perhutani, 1992).

Fig. 8.3. *Cunninghamia lanceolata* being grown on steep slopes in China where site preparation practices may cause excessive soil erosion and encourage grass and bamboo weeds (foreground).

Pinus elliottii *and* P. taeda *in the Southern USA*

As noted in Chapter 2, significant plantings of both *Pinus* species in the southern states began in the mid-1930s. With rotations usually 30 years or more, restocking commenced in the 1970s. In general, growth of the second rotation is variable – see examples in Evans (1999). A coordinated series of experiments throughout the USA is assessing the long-term impacts of a range of management practices on site productivity (Powers, 2001).

8.4.3 Are yields generally increasing?

There is evidence from a number of countries that productivity of planted forests is gradually increasing over time. This has been described as within-rotation Yield Class/Site Quality drift (Evans, 1999). Two recently observed phenomenon require comment.

Inaccuracy in predicted yield

For long rotation (>20 years) stands it is usual to estimate yield potential from an interim assessment of growth rate early in life and then to allocate a stand to a site quality class or index, or yield class. A change from predicted to final yield can readily occur where a crop has suffered check or other damage in the establishment phase or fertilizer application corrects a specific deficiency. However, there is some evidence for very long rotation (>40 years) planted forests in temperate countries that initial predictions of quality or yield class underestimate final outturn, i.e. the stands have grown better in later life than expected. Either the yield models used are now inappropriate or growing conditions are 'improving' in the sense of favouring tree growth. Across Europe the latter appears to be the case (Spiecker *et al.*, 1996; Cannell *et al.*, 1998) and is attributed to rises in atmospheric CO_2 and especially nitrogen input, increasing temperatures, better planting stock, and cessation of harmful practices such as litter raking (Kahle *et al.*, 2008).

However, as noted earlier, the opposite appears to be occurring with *T. grandis*. High initial site-quality estimates do not yield the expected outturn and figures are revised downward as the stands get older. And, as reported in Chapter 6, there is evidence from plots in tropical rainforests that growth rates may be slowing (Fox, 2007).

Correlation of quality (yield) class with time of planting

Closely related to the above is the observation that date of planting is often positively correlated with productivity, i.e. more recently planted stands are more productive than older ones, regardless of inherent site fertility. This shift is measurable and can be dramatic: see the example in Fig. 8.4 from Australia in Nambiar (1998). Fox *et al.* (2007) report that, for pine plantations in the Southern USA, the mean annual increment has more than doubled since 1952. In the UK, attempts to model productivity on the basis of site factors have often been forced to include planting date as a significant variable. Maximum mean annual increment of *Picea sitchensis* increased with planting date in successive decades by 1 m^3 ha/year (Worrell and Malcolm, 1990) and for *Pseudotsuga menziesii*, *Larix kaempferi* and *Pinus sylvestris* by 1.3, 1.6 and 0.5 m^3 ha/year respectively in each succeeding decade (Tyler *et al.*, 1996). This phenomenon suggests that some process is favouring present growing conditions over those in the past,

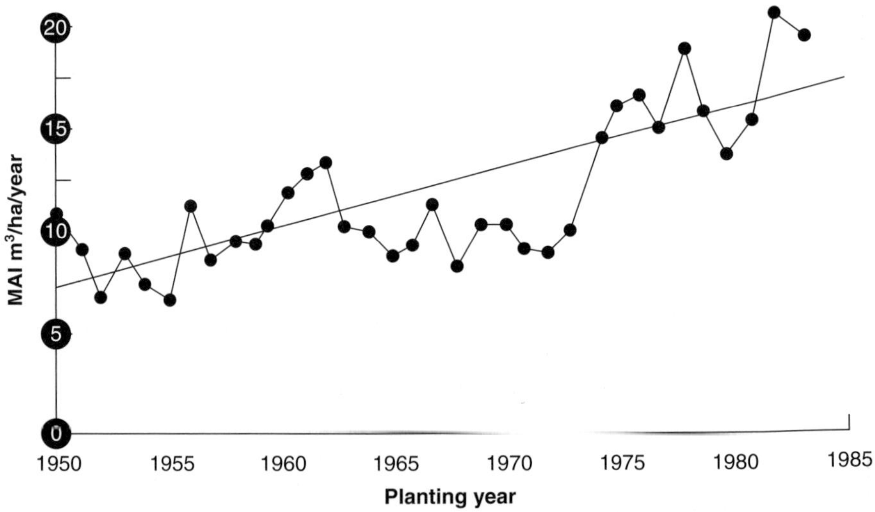

Fig. 8.4. Increasing productivity over time as genetics and silviculture improve: *Pinus radiata* in Australia (Nambiar, 1998) (reproduced by permission, E.K.S. Nambiar).

such as the impact of genetic and silvicultural improvements (and again cessation of harmful ones) and possibly the 'signature' of atmospheric changes mentioned above (Cannell, 2002). As a result, productivity of many planted forests in the UK may well increase owing to climate change.

The impact of these two related observations is that present forecasts of yields from planted forests may be conservative: the outlook is that yields appear to be increasing.

8.4.4 Some conclusions

Three main conclusions can be drawn from this and the previous section.

1. Planted forests and associated silvicultural practices do affect sites and under certain conditions may cause deterioration, but they do not appear inherently unsustainable. Care with harvesting, fire protection, conservation of organic matter and management of the weed environment are critical features to minimize nutrient loss and damage to soil conditions.
2. Measurements of yield in successive rotations of trees suggest that, so far, there is no widespread evidence that planted forests are unsustainable in the narrow sense. Where yield decline has been reported from one rotation to the next, poor silvicultural practices and operations appear to be largely responsible.
3. Evidence from several countries suggests that current rates of tree growth, including of planted forests, exceed those of 50 or 100 years ago.

8.5 Risks to Planted Forests and Questions of Ecological Stability

8.5.1 Pest and disease incidence in monocultures

A serious threat to planted forests can arise from a massive build-up of a pest or disease. It is much disputed whether monocultures are more susceptible to devastation from these causes. The broadly accepted ecological principle of stability is that the stability of a community and its constituent species is positively related to its diversity. Following this reasoning, ecologists and foresters have argued that compared with native forests, even-aged monocultures may possess fewer natural constraints on local tree pests and pathogens and thus increase risk of attack. Some evidence supports this, as in the early review by Gibson and Jones (1977), though these authors point out that increased susceptibility mostly arises from *conditions in planted forests* rather than because only one tree species is present.

The relative susceptibility of monocultures to organic damage is complex ecologically. For example, simply adding diversity by cultivating mixed crops may not offer much protection, since only small amounts of the right kind of diversity are needed to maintain stability (Way, 1966). Also the influence of diversity on the stability of insect populations, for example, depends on what population level is deemed acceptable. Often stable equilibrium levels are too damaging and artificially low populations are sought in order to keep below the damage threshold (Evans H. F., 2001). Pest control to maintain low levels is very different from that which achieves stability (Speight and Wainhouse, 1989), as every spray to every farmer's field testifies. These authors stress that artificially created diversity, i.e. mixed crops, does not necessarily improve ecological stability and is inferior to naturally occurring diversity, complexity of organization and structure.

The key point made by Wainhouse, in one of the most thorough recent reviews, is that: 'There seems no general evidence of a simple positive relationship between diversity and stability that is likely to be of value in predicting susceptibility to pest outbreaks in managed forest' (2005, p. 5). Nevertheless, he goes on to warn that a meta-analysis of 54 studies does show some evidence that insect abundance or damage can be higher in single- than in mixed-species stands. It is prudent, therefore, to spell out why planted forests consisting of one or two species is perceived to be at some risk.

1. Stands of one or two species offer an enormous food source and ideal habitat to any pest or pathogen species adapted to them. Food supply is a basic ecological determinant of population size and multiplicity of sites for breeding or infection favour rapid build-up.
2. Uniformity of species and proximity of trees, including branch contact above ground and root lesions in the soil, allow rapid colonization and spread of infection. Canker diseases that are splash-dispersed or mist-carried, and insects with small effective spread are favoured.
3. Narrow genetic base in planted forests, such as use of one provenance or no genetic variation (e.g. clones), limits variability in resistance to attack – a great risk that Wainhouse (2005) describes as 'unambiguous'.
4. Trees grow on one site for many years. This may allow a pest or disease to build up over time with little opportunity to destroy infection.
5. Exotic species, as are commonly used in planted forests, are often without the insect pests and pathogens that occur in their native habitat – the so-called pest and pathogen release hypothesis (Wingfield, 2004). This has undoubtedly contributed to the great success of *Eucalyptus* across the tropics, freed from numerous leaf-eating insects that occur in the Australian environment (Florence, 1996). Conversely, many natural agencies controlling pests and diseases are also missing and damage can be swift and uncontrolled. Planted forests of exotic species appear to experience an initial period of relative freedom from organic, damage but gradually this diminishes owing to accidental introductions of new pathogens and adaptation of native ones (Wingfield, 2004).
6. Extensive blocks of a single species may lead to some planting on sites to which it is ill-suited. As well as poorer growth, the greater stress this brings often renders trees more at risk from pest and disease outbreaks.

8.5.2 Examples of devastating outbreaks of diseases and pests affecting planted forests

1. *Dothistroma* needle blight of *Pinus radiata* in the East African highlands curtailed plantings after the 1950s, despite superior growth rates to *P. patula* and *Cupressus lusitanica*. In New Zealand, preventing this disease on *P. radiata* is one of the country's greatest forest protection expenditures, and in the UK it now threatens the future of *P. nigra* var. *calabrica*.

2. Diplodia (*Sphaerosis sapinea*) kills pines in Southern Africa if there is severe infection of tissue damaged by a violent hail storm.
3. In Europe, poplar canker (bacterial infection from *Xanthomonas populi*) and leaf diseases and rusts (*Marsonina* and *Melampsora* spp.) are so serious that laws restrict commercial use of poplars to a few relatively resistant species and certain clones of these species (Fig. 8.5).
4. In the 1980s outbreaks of the severely defoliating insect (psyllid) *Heteropsylla cubana* devastated *Leucaena leucocephala*, especially where Hawaiian hybrids were used, and this once widely planted nitrogen-fixing tree is now far less important.
5. Attempts to grow mahogany (*Swietenia* spp.) and other Meliaceae family species (*Cedrela*, *Khaya*, *Melia*, *Toona*, etc.) in planted forests often has been thwarted by stem deformation from the mahogany shoot borer *Hypsipyla* spp. (Fig. 8.6).
6. In the UK in the 1970s, the native moth *Panolis flammea* killed many young pole-stage stands of exotic *Pinus contorta* growing on deep peats in North Scotland.

Fig. 8.5. *Melampsora* spp. infection (poplar rust) causes premature defoliation and loss of yield. (Source FAO.)

Fig. 8.6. Shoot borer (*Hypsipyla* spp.) on African mahogany (*Khaya nyassica*).

These three examples of fungal pathogens and three of insect pests are merely illustrative of the scale and potential of threats. Further important examples will be found in Ciesla and Donaubauer (1994), FAO (2009), Wainhouse (2005) and Wingfield (2004). However, while these threats have prevented the planting of some species and impaired the productivity of others, overall they have not caused such widespread damage to seriously question the adoption of planted forests as a silvicultural practice. But, as Wingfield (2004) stresses, in the future increased investments will be needed to reduce impacts on planted forests, especially if they are of exotic species.

8.5.3 Will threats to planted forests increase in the future?

Environmental change

Many of the changes now observed in the composition of the Earth's atmosphere may threaten trees and forests, including planted forests. Rising greenhouse gas concentrations leading to climate change, and increasing atmospheric pollutants of particulates, nitrogen compounds and ozone will stress established planted forests and are likely to increase insect pest risk and diseases problems (Evans *et al.*, 2002; Lonsdale and Gibbs, 2002). The increasing incidence, scale and impact of extreme weather events (floods, droughts, hurricanes etc.) will need to be factored into risk assessment for planted forest investments in the future.

New pests and diseases will emerge

Pests and diseases new to science, new to a country or region, or new as a threat can be expected from:

- New hybrids, mutations and even species being identified, e.g. in the *Phytophthora* family of diseases (Brazier *et al.*, 2006) associated with 'sudden oak death' in America and Europe, death of *Alnus* in Europe, and *Eucalyptus* in Western Australia.
- New introductions from increasing global trade, especially from the massive expansion in shipment of live plant material (even semi-mature trees), e.g. *Cryphonectria* canker in *Eucalyptus* in South Africa, and *Cerambycid* (longhorn) beetles such as *Anoplophora* spp. from East Asia (in China it is severely damaging *Populus*) into Europe.
- Native pests adapting to introduced trees.

The challenge is how to respond to these biological threats. Traditional pest-risk analysis tends to be retrospective – organisms are on a list of threats because they are already known. The problem is how to prepare for the unknown. Recent history of pest and disease outbreaks in planted forests show that many were not predicted. Trade, including in germplasm and reproductive material as well as forest products, is likely to be critical. While unwanted introductions through imports of round timber, lumber, carcassing, pallets and dunnage have largely been eliminated through practices like debarking and enforcement of sterilizing treatments, phytosanitary certificates for live material are only as good as the pests and diseases looked for and which can be identified by visual inspection – not to mention risks from corrupt practices where defective material is passed as clean.

The outlook is to expect planted forests to suffer from biological threats: their composition renders them rather more at risk, and new threats are sure to arise. Vigilance in monitoring, assessment, reporting, control and research is clearly warranted.

8.5.4 Risks associated with specific practices

Some pest and disease problems of planted forests are exacerbated by the practices and operations used in their management, and not directly from growing one species of tree in a uniform way (monoculture).

Harvesting and other residues

Large amounts of wood residue from felling debris and the presence of stumps are favourable for colonization by insect pests and as sources of infection. There are many examples, several are cited in Evans (1999), but modification of silviculture or application of specific protection measures generally minimize such problems. Also, the way such residues are treated – burned, windrowed, scavenged or simply left – all impact the pest and disease environment.

Site and species selection

Mention has already been made that extensive planting of one species, whether indigenous or exotic, inevitably results in some areas where trees are ill-suited to the site and suffer stress. This may occur where large monospecific blocks are planted or where exotics are used extensively before sufficient experience has been gained over a whole rotation, e.g. *Acacia mangium* in Malaysia and Indonesia and the discovery of widespread heart rot.

Thinning and pruning damage

Thinning operations can damage remaining trees and provide infection courts for diseases and, in the case of Fomes infection (*Heterobasidion annosum*), as stumps are colonized and through root lesions, lead to death of adjacent trees. Delayed thinning, ragged pruning and poor hygiene can also increase risk to remaining trees (several examples are cited in Evans and Turnbull (2004)), and while none seriously threatens the sustainability of planted forests, they do emphasize the need for good husbandry.

8.5.5 Are planted forests more prone to droughts, storms and fires?

Drought

Drought is one stress likely to be encountered by planting large uniform blocks with no account of soil depth and other physical characteristics. In many countries worsening droughts are expected as a consequence of climate change and species matched to sites based on past climatic patterns may be at risk in the future, e.g. *Fagus sylvatica* in Southern England (Broadmeadow and Ray, 2005). And drought stress, along with inadequate thinning, renders *P. radiata* especially prone to *Sirex* wood wasp infestation.

Storms

The uniformity of planted stands may increase the risk of hurricane and storm damage because (a) trees may be planted in locations that increase their susceptibility, such as exposed sites or on soils that do not permit deep rooting, and (b) trees in dense stands are architecturally more prone to wind snap and their root development is more restricted, making them unstable. While minimizing hurricane damage in the tropics can, for example, be helped by planting wind-firm species such as *Cordia alliodora* or choosing *Pinus caribaea* var. *bahamensis* over *P. oocarpa* (Evans and Turnbull, 2004), recourse to more costly site preparation that aids rooting depth and no-thinning strategies are needed in the upland UK to minimize this wind damage in extensive planted forests of *Picea sitchensis*.

8.5.6 Forest fires

Most forest fires that damage planted forests are caused by arson and only a few by lightning or encroachment of fires from neighbouring land. Jurvélius (2004) cites 95% of fires as caused by human activities. While there are examples of frequent fires preventing successful establishment of planted forests, the issue is more that of relations with the local community than an inherent shortcoming with planted forests *per se*. Of course, young stands are especially prone to ground fires and older ones can have massive fuel build-up, creating hazardous conditions, both of which render planted forests susceptible.

The threat, like that of droughts and storms, is more a question of a risk to an investment – all planted forests are an investment to a greater or lesser extent – and due care to address the level of threat will have to be taken. This may increase as climate change and weather disturbances such as El Niño events exacerbate droughts.

8.6 Enhancing Production and Other Benefits from Planted Forests

The steady transition from exploitation of native forests to dependence on planted forests is following the path of agriculture. Many of the same biological means to enhance yield are available. They are outlined here only briefly and are summarized for their potential to increase productivity, to alleviate threats and to suggest what best practice is and as detailed in FAO (2006b).

Also outlined are other interventions in a stand of trees or area of planting to aid delivery of non-market benefits such as biodiversity. The wider issues concerning such roles for planted forests were considered in Chapter 6.

8.6.1 Genetic improvement

There is only one opportunity per rotation to change a forest crop. Change in species, seed origin, use of new clones, use of genetically improved seed and, in the future, genetically modified trees all offer the prospect of better yields, greater resistance to pest and disease threats, improved wood quality, and coping with climate change.

Species change

There are surprisingly few examples of wholesale species change from one rotation to the next, which suggests that in most cases foresters have been good silviculturists. Five examples, each of which led to increased productivity, are cited in Evans (1999).

Better seed origins, provenances and land races

The impact of such genetic improvements will affect yield and outturn directly and indirectly through better survival and greater suitability to the site, which may lead to increased vigour, improved stem form and perhaps greater pest and disease resistance. Countless studies affirm the benefits of careful investment in this phase of tree improvement to the point that their use may be mandatory in some circumstance. For example, in the EU, planting what are categorized commercial tree species must only be with seed from approved sources, such as registered seed stands.

Clonal planting stock

Some of the world's most productive planted forests use clonal material (Fig. 6.1), especially with *Eucalyptus*, *Salix* and *Populus*. For conifers, micropropagation and tissue culture is allowing exploitation of this technology. It is clear that both the potential productivity and the uniformity of product make this silviculture attractive. Although clonal forestry has a narrow genetic base, careful management of clone numbers, continual recruitment of new ones, the way they are interplanted in a stand or as a patchwork across the landscape and from one year to the next can minimize pest and disease problems. Typically 10–30 unrelated but well-tested clones are planted operationally to provide security against catastrophic failure (White, 2004). In Brazil companies prefer to have a mosaic of 10–20 clones in any given area and no more than 40 ha of any one clone (McNabb, 2004).

Tree breeding

Through an array of selection, crossing and propagating techniques, traits can be favoured that may improve vigour, stem and wood quality, pest and disease resistance and other parameters such as frost tolerance. Examples of yield gains were cited earlier (p.62). Genetic tree improvement offers the best assurance of sustained and improved yields from planted forests in the medium- and long-term (Fig. 8.7) and is considered to offer better economic returns than other alternatives in forestry (Kanowski and Borralho, 2004).

Genetically modified trees

Use of genetically modified (GM) trees is still in its infancy. The technology offers herbicide tolerance, pest and disease resistance, modified wood quality, such as reduced cell wall lignin in pulpwood, greater cold or drought tolerance, and reproductive sterility (Burdon, 2004). There are no extensively planted forests of GM trees, except in China, where Marchadier and Sigaud (2005) report that use of insect-resistant GM poplars is approved. Public acceptance of such technology is an open question in some countries (Evans, 2004) and in the USA, for example, permission is needed to plant GM trees, especially if they are to be allowed to flower.

Responding to environmental change

Genetic improvement or changes offer a means of responding to climate change. For genetics to fulfil this role, an active breeding programme is necessary and must be based on a wide genetic diversity. Part of the programme should not only be selection and multiplication, but also deliberate maintenance of a broad genetic base. This may require new collections of material from across the whole

Fig. 8.7. Seed collection from a superior stand of *Pinus caribaea* in KwaZulu Natal, South Africa.

of a species' natural range, and selection and screening processes to anticipate particular climate change scenarios (Burdon, 2004; Karnosky and Thakur, 2004).

8.6.2 Silvicultural interventions likely to increase yields

Greater understanding of tree and stand physiology can be expected to deliver incremental gains in yield per hectare, i.e. intensify production, notably:

- Manipulation of stocking levels to achieve greater output of total fibre, or a particular product such as high-quality lumber (Savill and Evans, 2004).
- In a few situations mixed-species crops to help tree stability and possibly lower pest and disease threats (FAO, 1992).
- Optimizing rotation length to meet volume or technical specifications and targets.

- Targeted fertilizer application to correct known deficiencies, e.g. phosphate and boron in much of the tropics, and redress imbalances, e.g. nitrogen where litter breakdown is slow, and at appropriate stages in stand development – at planting, pole-stage etc.
- Greater operational use of fertilizers in regions with poor soils as is now accepted as standard in countries such as Brazil (Gonçalves *et al.*, 2004b; McNabb, 2004).
- Improved anticipation of catastrophic damage such as storms and pest/ disease outbreaks, and salvage during restoration.
- Advances in detection and suppression of forest fires.
- Intensifying extraction to improve volumes harvested per hectare, including for fuel and renewable energy usage, but with the long-term threat of increasing risk of site degradation (Moffat *et al.*, 2006).
- Effective vegetation management (weed control) strategies (Wagner *et al.*, 2006).
- Holisitic approaches to harvesting and re-establishment so that future yield is not sacrificed for short-term savings, e.g. delayed replanting to allow weevil numbers to fall, careful re-use of extraction routes to minimize soil compaction and erosion, and minimizing soil disturbance to conserve organic matter and reduce weed problems (Gonçalves *et al.*, 2004b).

8.6.3 Silvicultural interventions to meet other objectives with possible yield penalties

For reasons usually of amenity or enhancing biodiversity, certain sivicultural actions may reduce productivity in terms of forest products.

- Silvicultural systems that maintain forest cover at all times – continuous cover forestry practices – such as shelterwood and selection systems are likely to be neutral to slightly negative in production terms, while yielding gains in tree quality, aesthetics and probably biodiversity value. Modelling suggests they may be the best combination of high-wood yield and carbon storage (Thornley, 2007).
- Extended rotations for amenity purposes near built-up areas or to create 'old-growth' conditions in planted stands (Humphrey, 2005).
- Deliberate exclusion of land (or withdrawal from an existing planted forest estate by not replanting) for reasons of wildlife conservation, archaeological or cultural interest.
- Reduced pesticide use – from its already low level – which may increase damage levels, though not necessarily if more precision in application can be achieved (Willoughby, 2007), or by switching to organic products, which may be less efficacious (Clay *et al.*, 2005).
- Policies that switch planting from exotic to native species to restore more native forest conditions – in Europe often from conifers to broadleaved species (Harmer *et al.*, 2005) as noted in Chapters 2 and 6 – will often have a production penalty.

8.6.4 Enhancing biodiversity within the planted forest estate

Much has already been said in Chapter 6 concerning the roles planted forests can play in relation to biodiversity, but there are beneficial interventions possible in almost any planted forests with little loss of production (Peterken, 1996; Carnus *et al.*, 2006).

Conserve existing features

When laying out and establishing new planted forests, conserving native woodlands and wetlands, creating buffer zones next to watercourses and lakes, avoiding archaeologically or culturally important sites and similar actions will minimize damage and assist conservation.

Restore features

When regenerating areas planted in the past, opportunities to restore significant features that perhaps were overlooked previously or not sufficiently valued should be reviewed, e.g. Fig. 2.5. For example, wetlands that were once seen as candidates for site drainage may best be restored by not replanting, and the blocking of drains and ditches (Gill and McIntosh, 2001) – growth may be poor anyway and the wildlife gain far more significant.

Create diversity of structure as well as species

While emphasis is often placed on using native tree species and planting a mix of species to enhance biodiversity, it is clear that ensuring diverse structure in a forest is also important. Uniform stands of native or exotic species can be equally unattractive to wildlife. Encourage structural diversity through different age-classes, creating glades, maintaining open tracks and firebreaks, wider spacings, earlier as well as timely thinnings, and varying regeneration (felling coupes) to break up large uniform blocks.

Extend rotations and conserve truly ancient trees

The concept of 'old growth' and the recognition that mature and over-mature trees of native and exotic species provide habitats for specialist insects and fungi and encourage native ground flora development (Er and Innes, 2003; Humphrey, 2005) indicate that a small proportion of a planted forest should be retained for this purpose. Such trees will often become a valued amenity (Fig. 8.8).

Dead wood habitat

Particularly in short-rotation crops, little dead wood accumulates on the forest floor. Indeed, sometimes such debris can be a threat to forest hygiene or a fire risk. Nevertheless, leaving snags (standing but dead tree trunks) and woody material on the forest floor, where possible, adds ecological niches.

Fig. 8.8. Ancient *Quercus robur*, planted about 270 years ago and previously pollarded, now of immense value for wildlife and biodiversity.

8.7 Threats Planted Forests Pose

Several issues relating to impacts of planted forests on hydrology, soil conditions and biodiversity conservation, as well as the critically important social dimension, were considered earlier. Here two issues are addressed.

- Does planting forests, in particular the use of exotics, threaten other ecosystems owing to the planted species becoming a weed and 'invasive'?
- Can silvicultural operations themselves introduce biological threats – e.g. a centralized nursery or the nursery trade spreading infected germplasm?

8.7.1 Planted species as invasives

The risk of planted trees of exotic species becoming weeds is a possibility. Where good natural regeneration occurs, that is there is plentiful fertile seed and ground

Fig. 8.9. Invasive *Acacia longifolia* in the Clifton area of Cape Town, South Africa. Two biological control agents, a wasp and a rust fungus, are being used to attempt control. (Source: J. L. Innes.)

conditions are suitable, unwanted regeneration can become invasive, e.g. *Tsuga heterophylla* in the UK. As concern has grown over the past decade for the protection of indigenous flora, there has been strong pressure to choose species for planting with little or no such weed or invasive potential (Hughes, 1994). Serious problems with aggressive woody weeds such as *Acacia nilotica* in Australia, *A. melanoxylon* and *A. saligna* in South Africa (Fig. 8.9), several *Prosopis* spp. in India, Pakistan, Kenya, much of the African Sahel and the USA, and *Melaleuca quinquenervia* in Florida, USA illustrate the potential dangers of some exotic trees. Leguminous trees and shrubs with hard seeds that persist in the soil have a relatively high risk. Many, such as the widely planted *Acacia mangium*, seed at an early age and, as pioneers, readily colonize bare ground, especially if created by fire. In general *Eucalyptus* species are rarely invasive, though *Pinus* can be more so (e.g. *P. patula* in the Eastern Highlands of Zimbabwe, and *P. contorta* in South Island of New Zealand). In a review of invasiveness of conifers, Rejmánek and Richardson (2003) include several widely planted species as amongst the most invasive. In addition to the two *Pinus* species just cited, they single out as serious invaders *P. elliottii*, *P. halepensis*, *P. kesiya*, *P. pinaster*, *P. radiata*, *P. strobus*, *P. taeda* and *Picea sitchensis*, and *Pseudotsuga menziesii*.

In many places the extent of the weediness (invasive) problem is only just being realized (Evans and Turnbull, 2004). However, Goodall and Klein (2000) already list 18 alien tree species in South Africa exhibiting invasive tendencies. Clearly a risk assessment of any new introduction and careful monitoring during evaluation is called for. For species that have already been introduced, the lists of recommended species must be closely scrutinized to avoid weedy species or care taken when planting to minimize the chance of escape. However, sometimes a fear or claim of invasiveness diminishes as further research revises initial concerns, e.g. *Acer pseudoplatanus* colonizing *Fraxinus* woodland in the UK.

8.7.2 Threats from silvicultural practices

Some operations and practices associated with planted forests could themselves pose threats to such forest. Distributing planting stock from centralized nurseries can unwittingly but very effectively spread pests and pathogens. Similarly, timber movements of untreated logs can act as vectors. Overzealous thinning and pruning or poor forest hygiene can increase threats. Inadequate monitoring of forest health can allow serious threats to go unrecognized. In the UK, *Dendroctonus micans* was not identified until 9 years after its introduction, by which time it had established as a major pest.

8.8 Trees on Farms and Non-forest Plantings

Much in this chapter applies to trees on farms and, generally, to planted trees outside the forest. They can of course spread the risks, e.g. the problems a decade ago with *Leucaena leucocephala* and psillid damage. However, the general expectation is that such plantings will increase and appraisal does not suggest that they are under any particular threat or pose a threat themselves (Sinclair, 2004).

9 Summary and Conclusions

J. EVANS

9.1 Planted Forests at a Critical Point

From time to time there comes a tipping point when an occasion or set of circumstances clearly set the agenda for the future. Such a critical point appears to have been reached with the ascendancy of planted over natural forests for supplying many industrial commodities. Globally, 70% of forest products are sourced from 7% of the world's forest, of which the latter are predominantly planted or have a planted component. 'Planted forest' includes all of what is generally understood as 'plantations' or 'forest plantation', but also incorporates other forest types originating largely or wholly from tree planting.

In asserting this view – that planted forests are playing a role far in excess of what their actual area might suggest – limitations are recognized. They do not and cannot substitute natural forest formations: they are not an alternative but are complementary. They may help ease some pressures on natural forests, but that is unlikely to be their major role, desirable as this might be. Planted forests, in all their variety, offer major opportunities but are no panacea to the ills that beset the world's forests at large, namely, destruction and deforestation, the loss of ecosystems and environmental services, and perhaps most important of all, the loss of somewhere to live for many people too often on the fringe of society. Tree planting and planted forests have a role to play and are part of the solution to these ills.

This chapter draws together the key points concerning the status of planted forests in the form of collating extracts from the preceding chapters.

9.1.1 Some conclusions from the history of planted forests

The history of planted forests reveals issues that have arisen and draws out pointers for the future.

- Tree planting has a long history: it is a very old practice.
- Reasons for planting forests were frequently prompted by local or regional shortages of wood, fuel and lumber.
- While afforestation – creating new forest by planting – has always featured, much early use of tree planting was to augment inadequate or missing natural regeneration. Tree planting ensured satisfactory restocking of existing forest.
- Use of exotics has resulted in some of the world's most productive planted forests.
- Concerns over sustainability, and more generally the use of monocultures, go back at least 100 years, but have not seriously challenged silvicultural and management practices associated with planted forests.
- While much tree planting has had production objectives, planting for soil protection, assisting conservation and other multipurpose roles have been commonplace.
- There has been a long history of tree planting outside the forest, including agroforestry, to provide goods and services and to combine food and tree growing in socially acceptable ways.
- While some planted forest practices have intensified to increase productivity per hectare, there has also been recognition that management over whole forests must be environmentally sensitive in the interests of landscape, amenity, conservation and related imperatives.
- Stakeholder involvement to influence planted forest development and to engage in tree and forest planting itself is crucial.

9.2 Refinement of Definitions

The history of planted forests reveals the extent to which recourse has been made to tree planting. As such forests have matured and been regenerated, this history was often overlooked or forgotten. Similarly, planting of native species often quickly led to quasi semi-natural forest conditions far removed from the common perception of a plantation. Adaptation of the FRA 2005 categories by the Global Planted Forest Thematic Study addresses this broad spectrum (FAO, 2006c).

The logic is that the planted component of semi-natural forest, with intensive silvicultural treatments, is not materially different from forest plantations. The only distinction is that it is of native species and continues the character of the previous forest and such planted forest is, in fact, very common.

9.2.1 Implications

The expanded concept of 'planted forests' brings together two broad kinds of forest that were formerly considered separately: plantations and the planted component of semi-natural forests.

Because of this new grouping and because planted forests reflect a higher social, environmental and economic importance than their area would suggest, FAO embarked on a Global Planted Forest Thematic Study (FAO, 2006c). This in-depth examination of the whole planted forests subset included both productive and protective functions, while still distinguishing between them, with the grand aim of providing a realistic picture of the status and trends of wood, fibre, biomass, non-wood forest products and services from planted forests.

9.2.2 Impact of planted forests

The significance of the planted forests subset of the world's forests is striking.

- The estimated area of planted forests, 271 million ha, amounting to 7% of the world's forests, points to a more significant resource than hitherto recognized.
- Planted forests continue to expand, and will likely play an even greater role in the future.
- Distinguishing between planted forests as a component of semi-natural forests and those traditionally labelled forest plantations has helped to reveal the extent of plantation silviculture in forest management and practice globally.
- Policy-making and planning, as well as investing in protection (fire, pests and disease), research and maintenance, must recognize the fact of and trend towards planted forests.
- Data on yields and end-uses suggest that planted forests contribute massively to industrial wood and fibre supply (potentially two-thirds), which in the past was underestimated.
- Distinguishing between functions of planted forest – production and protection – reveals the surprisingly substantial investment in environmental protection.
- The impact of large volumes of limited species and size classes on markets in the near future must be considered by planners, while technologists may have to develop ways of utilizing large quantities of small-dimension logs.
- The impact on wood supply, as well as on provision of environmental services, of increased private ownership of planted forests, including by smallholders, is a critical shift and raises some uncertainties about the continuity of supply that may need to be addressed through policy-related measures.

9.3 Multiple Roles of Planted Forests

Planted trees and planted forests serve many roles and functions. There are many opportunities to take advantage of management and silviculture with planted forests to balance the social, cultural, environmental and economic trade-offs. Just as it is poor forest stewardship to clear native forest simply to provide somewhere to plant trees, it needs to be recognized that there is plenty of degraded, waste or cut-over land entirely suitable for planted forest. It also needs affirming that planted trees and forests are able to serve several roles at once. It is harnessing trees for their wood, fibre and non-wood forest products, their many influences (environmental and social), and their interaction in the ecosystem for the good of all that is the aim.

9.3.1 The future fibre basket

The new planted forests data and the past underestimation of the role planted forests play are changing perceptions concerning supply of forest products. In one generation a sea change has occurred. The optimism about 'man-made' forest and the speculation concerning planting programmes 40 years ago at the Symposium on Man-made Forests and their Industrial Importance (FAO, 1967) have, it can be argued, not only been realized but exceeded.

Overall, fibre supply from planted forest is set to increase dramatically and eliminate any lingering spectre of wood shortage globally, if not always locally. Not only will the resource of planted forests largely meet current levels of demand for industrial wood, but, in the medium term, surpluses are possible that can make inroads into and substitute use of non-renewable construction materials that are far more energy-intensive – cement, steel and aluminium (Bowyer, 2004). Such surpluses are highly likely if significant investment occurs in carbon afforestation and reforestation as a climate change mitigation strategy. Not only will much virgin wood be grown but, ultimately, its very best use is a win–win, both to substitute for other materials as a renewable low energy-consuming alternative and in ways that have long in-use life, and so prolong carbon storage. Consumer preference is beginning to place a premium on environmental and social justification in product use. Solid wood and reconstituted wood products may regain market share in the construction industry.

Planted forests are becoming the world's industrial feedstock for wood products. This will require similar inputs as farm crops of high-quality germplasm, site/species matching and a relatively high intensity of silviculture – establishment, protection, management, harvesting and regeneration. Such forests will represent one branch of an emerging dichotomy, namely, intensive cropping for industrial end-uses in contrast to less intensive management for many non-industrial uses.

9.3.2 Energy feedstock

A key issue pointing to massively increased uptake of bioenergy, including bioenergy from planted forests, is their use for gas or liquid fuel production. Wood pyrolysis and wood distillation, the latter for methyl and, particularly, ethyl alcohol (Zerbe, 2004; FAO, 2008a) are becoming attractive as sustainable energy alternatives. Development of cost-lowering technologies for the processes could bring these fuels on stream at economically competitive rates from planted forests and without some of the downside – displacement of food production, massive clearances of native forests – associated with fuels derived from cereal grain, sugar or vegetable oils.

9.3.3 Will forests be planted to store carbon?

Trees accumulate carbon during their life and can often enrich soil surface and sub-soil carbon stocks with recycling of organic matter, their root systems and formation of peat. Thus appropriate afforestation and reforestation are seen as weapons to slow the rise in atmospheric carbon dioxide and in turn help mitigate global warming. However, exaggerated claims have been made and it has even been argued that because planted forests mostly grow faster than native forests, it is best to clear them and plant trees. Such action will rarely be carbon positive, ecologically sound or socially justified. Indeed, the best single contribution the forest sector can make to mitigate climate change is to prevent deforestation (Kirschbaum, 2003; Innes, 2004). Even forest degradation that lowers carbon stocks is increasingly seen as a concern. Overall, however, there remains an expectation of many new planted forests specifically to sequester carbon.

9.4 Some Social Roles of Planted Forests

9.4.1 Poverty reduction and sustainable livelihoods

The Millennium Development Goal of eradicating poverty naturally raises the question: what role do planted forests play? Being overwhelmingly a rural development – while recognizing the importance or urban and peri-urban plantings – and because many poor people, including subsistence farmers and many landless and refugees, are rural dwellers, can planted forests alleviate this privation? Conversely, and to put it bluntly, can planted forests also exacerbate poverty by, for example, displacing people from their land or disrupting local socio-economic patterns? The issue also bears on the urban poor since many people who live in towns and cities rely on wood for fuel and building materials.

Large-scale investment in planted forests can undoubtedly benefit many poor people, and with a favourable and creative policy environment can contribute to alleviating poverty. But as Turnbull (2007) concludes concerning socio-economic impacts of China's extensive eucalypt afforestation programme: 'although plantation development has contributed significantly to poverty alleviation, it is probable that greater benefits accrue to higher-income groups'. It is clear that investors need to recognize and encourage activities with local communities wherever practical to do so.

There is an abundance of literature on rural development forestry and, as pointed out, community and social forestry have by no means always been successful. Medium-scale developments of planted forests have created many new resources, but are no panacea for all the ills of the poor. And, as stressed earlier, for planted trees and forests to help, there must be full and sufficient engagement with, and participation of, all stakeholders, but crucially the 'beneficiaries' themselves (Mayers and Bass, 2004). Foisting solutions on people rarely works.

9.4.2 Can planted forests slow deforestation?

This question is more of a synthesis of other roles and is addressed specifically because it is often asked. It is shaped in terms of a role planted forests could play for the benefit of society, rather than a more specific purpose. The issue tends to focus on the tropics, but in many countries where forest cover has doubled or tripled in the last 100 years through afforestation (Denmark, Ireland, the UK), or where planted forests overwhelmingly provide the industrial feedstock for wood products (Chile, New Zealand, South Africa and even the USA), it has become possible to divert pressure away from, and so conserve remaining areas of, native or semi-natural woodland and forest. Does this have a wider relevance?

Few would disagree with Mather's (1993b, p. 207) conclusion that 'planting trees would in itself not save the forests, but it is a good beginning' and it is 'certainly difficult to see the global forest being saved if trees are *not* planted. It does not itself offer a solution or a "technical fix" to the problem of shrinking forest resources'. Planting forest is not going to be a primary tool to slow deforestation, just one more help. But, importantly, a help too in public perception in the West that says: using wood destroys rainforest when in fact most paper and wood products come from sustainably managed planted forest. Greater awareness of this could be crucial, and sharpen the focus on the real drivers of deforestation.

Planted forests have a part to play in slowing deforestation, but will rarely be the principal means.

9.5 Environmental Services

9.5.1 Can planted forests help to mitigate climate change?

The role of planted forests for bioenergy and carbon storage were noted, but it is the huge threat climate change poses that may promote their development. If planted forests are fast satisfying world demand for timber and if new planted forests for bioenergy and carbon are established, their greatest contribution could be to provide feedstock for renewable fuels and for alternative construction materials to substitute for energy-intensive aluminium, concrete and steel and do so in a use which often stores carbon for decades. To the extent that these developments occur, planted forests are a win-win.

9.5.2 Protective afforestation

The role trees and forests play in protecting the environment should not be underestimated – hence the broad classification of forests into productive and protective in FRA 2005. Forests are a key element in the landscape. They provide a natural buffer that helps to maintain ecological balance, and supply raw materials to communities as well as provide protective services. Forest practices need to ensure the health and vitality of forests if they are to be managed sustainably and fulfil their protective role, especially protection of water resources, soil, the buffering local climate through control of wind, and other conservation and recreational functions (Göttle and Sene, 1997).

9.5.3 Do planted forests adversely affect hydrology?

When water from a catchment is used for drinking, irrigation or generation of hydroelectric power, management of the watershed is critical since it can affect both the quality and quantity of water supplied. Changes in vegetative cover by deforestation or tree planting have significant effects on the hydrology of a watershed. What these effects are and their scale can be contentious. Calder (2002, p.38) suggests that claims that forests increase runoff, regulate flows, reduce erosion, reduce floods and improve water quality are 'either exaggerated or untenable'. He argues that foresters and hydrologists often have different perceptions regarding the hydrological role of forests, with hydrologists concerned that trees intercept more rain during wet periods, and because of deeper root systems deplete groundwater by transpiring more water in dry periods (Calder, 1996). He also suggests that it is unrealistic to attempt to generalize forest impacts as they affect extreme flows. Also, some negative impacts related to floods and erosion may be more associated with forest management operations such as logging, site preparation and roading, rather than the presence or absence of the forests themselves. Non-hydrologists have sometimes over-simplified the impacts of forests in water catchments, especially when attempts are made to generalize from reports of events or research that may be very site-specific.

Because water availability is one of the great issues of the present century and will be exacerbated by climate change, the relationship between forests, and especially planted forests, and water supply is crucial. Good practice integrates forest with other land uses and ensures their sustainable management (FAO, 2008b). Planting trees is only ever one component of better land-use practices in catchments. The two key conclusions are:

1. To minimize erosion and sediment release it is best to retain undisturbed native forest and vegetation, including grassland. Where tree planting is carried out, it is essential to limit soil disturbance, and to encourage understorey vegetation and a well-developed litter layer.

2. Afforestation of land where planted trees replace grassland, scrub or degraded land will usually reduce water yield, peak and base flow rates, though where there is good infiltration base flows may be prolonged. Small to medium-sized stormflow events and consequent flooding may be reduced.

9.6 Ecological Roles

9.6.1 Can planted forests protect threatened natural ecosystems?

The concept of an 'island' of undisturbed native forest protected by a collar of planted forests to act as a buffer is appealing. It suggests deliberate attempts at conservation. Such buffer zones certainly bring micro-climatic amelioration in the transition from non-forest to native forest and may even assist wildlife conservation at the forest edge (Boston and Sessions, 2006; Denyer *et al.*, 2006). A more common situation is fragmentation of native forest and the impact that planting trees and planted forests can have on aiding secondary succession in gaps or re-colonization of native bird and animal species in a landscape mosaic.

What is clear is that using planted forests to help conserve threatened ecosystems can be significant but will never be a major role. What is emerging as a major role is conservation by default as well as by design. Application of best practice guidelines avoids damaging important ecosystems in areas of planted forest, such as wetlands, fragments of native forests, and even individual ancient trees.

In all management of planted forests, biodiversity is enhanced by:

- Conserving existing natural features – fragments of native forest (in their own right and as seed sources), wetlands and other special habitats – both by avoiding planting them and even reinstating them where possible in existing planted forests.
- Retaining individual ancient, unusual or very large trees.
- Encouraging development of trees and stands of all ages and conditions, including extending rotation length to simulate 'old growth' conditions.
- Avoiding uniformity and, in particular, diversifying structure – edge habitat, open areas, tall trees, new growth.
- Providing areas of light and shade – glades, open rides, firebreaks.
- Leaving some dead trees standing – snags – and creating deadwood piles.
- Resuming traditional practices, where appropriate, such as coppicing and pollarding.
- Minimizing soil disturbance and loss of soil seedbank.
- Managing 'new' habitat such as fire dams with a view to their wetland potential.
- Choosing native species for planting in preference to exotic species.

9.6.2 Planting trees and forest for ecosystem rehabilitation and restoration

Globally, the damage to native forest formations has been both a decline in their extent (deforestation) and a decline in their condition (forest degradation), leading to losses in terms of wood and non-wood products and of environmental services. In particular, biodiversity has been greatly reduced and few areas can recover unaided. Tree planting can assist recovery and help to rebuild indigenous forest ecosystems, a process known generically as 'forest restoration'. It is a role that is increasing in importance and there are both tropical and temperate examples where the act of planting reintroduces the woodland environment to initiate the process of recovery.

9.6.3 Reclamation of special kinds of sites

Planting trees as agents of rehabilitation or bioremediation of degraded sites, such as mining waste, salt-affected soils or tailings, as the first step of revegetation has a long history. It aims 'to recover productivity of a degraded site mostly using exotic tree species' but 'the original biodiversity is not recovered although the protective function and many of the ecological services may be re-established' (Lamb, 2001). Species with exceptional physiological tolerances to improve site conditions and initiate soil-forming processes, such as *Acacia*, *Alnus*, *Betula*, *Eucalyptus*, *Pinus*, *Salix* and other pioneers, are frequently employed.

9.6.4 Planted forests and conservation of genetic resources

The direct role planted forests play in conservation of genetic resources is limited, and is rarely if ever a principal object of management. The obvious exceptions are arboreta and similar tree collections. However, countless seed collections, followed by tree introductions, species trials and tree breeding programmes at the heart of planted forest silviculture, have created banks of germplasm which, today, may sometimes be of genuine importance for ex-situ conservation of genetic resources. This arises from loss of native forest where the species occurs, e.g. many tropical *Pinus* areas in Central America, or where seed collections in remote areas coincided with good fruiting of otherwise difficult to obtain species.

9.7 Policy, Institutional and Ownership Issues

9.7.1 Introduction

The evidence presented here of the emerging dominance of planted forests for meeting much of the world's production needs points to change also in policy, legal and institutional frameworks. There are many issues, but compared with most native forest they are less intractable since planted forests usually have distinct boundaries, will have required investment to create and to care for them, and are managed in accordance with some specific objective(s): their artificial nature makes for less complexity.

9.7.2 Who will own planted forests?

In Chapter 2 the historical trend in ownership from public to private in the last 30 years was noted. Data from the Global Planted Forest Thematic Study (FAO, 2006c) show that for planted forests for production, public ownership fell from 70 to 50% in the last 15 years, but where the protective function was uppermost, the bulk remained in public ownership. And, for both function types, there is an additional trend to increased smallholder ownership.

9.7.3 Overall outlook in planted forests investment

Massive 'investments' have been made in planted and managed native forests by institutional and high net wealth investors over the last decade. However, almost all have gone into existing planted and managed native forests, and have not led to expanding the area of planted forests, with a few, very minor exceptions in Latin America.

Historically, the expansion of planted forests correlates directly with consistent government support coupled with affordable land purchase/rent costs and identified markets for wood products. The alignment of these positive factors in several countries which occurred from 1960 to 2000 is no longer so obvious. Nevertheless, several key conclusions can be drawn.

- Over the past 50 years, most countries that have expanded their sustainable, commercial planted forest estate have had in common: stable governments, strong security of land tenure and an independent judiciary to protect investor rights. Almost all provided generous tax relief or subsidies.
- The most effective way international institutions can ensure expansion of private planted forest investments in developing countries is to assist governments to improve security of land tenure, and to protect investors by independent legal systems.
- The ownership of planted forests in many countries in North America, Oceania and Latin America has been transformed in the last 30 years, and particularly in the last 5–10 years. Governments and integrated forest product ownership no longer dominate; institutional and private equity funds are now big players. Additional investors and funding sources are entering the sector.
- There are many initiatives from governments and political blocs to increase substantially the area of planted forests, largely driven by the global warming and carbon sequestration issue. However, the mechanisms to secure the necessary land and specific funding sources remain unclear.
- Planted forest expansion will face increasing competition, in both developed and developing countries, from food production and the emerging biofuels industry.

9.7.4 Smallholder ownership and the issue of scale

The proliferation of smallholder ownership of planted forest can bring problems of access to market intelligence, poor technical knowledge and support, inefficiency in operations – or simply poorer management as well as increased bureaucracy. Traditionally extension services have met these needs, but the trend in small ownership will require changes in policy and institutional frameworks such as group certification, facilitating cooperation in management, protection and silviculture, simplification of regulations etc.

9.7.5 Stakeholder issues

Trees and forests by their very nature – their scale and influences – attract interest and concern far beyond that of their owners. Many stakeholders, from owners of adjacent land deprived of light or moisture by trees to those simply critical of forests that are planted, feel threatened. Many stakeholders gain direct employment or possess rights they can exercise. Many stakeholders are investors or derive income from planted forests. And the public at large see their landscape, environment, even their water supply affected by planted forests to name some non-market impacts. All these and other interests are now recognized. Ensuring engagement with stakeholders in ways that are just and sound is at the heart of responsible management of planted forests and are spelt out in FAO's voluntary guidelines (FAO, 2006b). Compliance with them is reviewed with care in the certification process.

9.7.6 Regulation and certification

The simpler the type of forest and the more manageable the scale of planted forest renders it amenable to certification. The vast majority of currently certified forest areas (both tropical and non-tropical) are for production (Durst *et al.*, 2006), and most fall within the definition of planted forests. Today, certification is seen as conditional for public acceptance that forests are managed sustainably (Higman *et al.*, 2005). That independent certification will expand further and impact most planted forests appears a given at the present time: it is becoming the principal tool for ensuring high standards of development, management and operational practice.

9.7.7 Illegal practices

Compared with native forest, planted forests usually suffer less from illegal practices. Ownership, investment and economic activity, and higher levels of management, all bring a greater degree of care and less opportunity for some types of crime, particularly that facilitated by remoteness and distance from supervision.

9.8 Sustainable Management and Silviculture

Many biological and environmental factors can enhance or threaten the outlook for planted forests. They can often be anticipated and so guide the direction of planted forests' management. Because the act of planting forest deliberately brings greater uniformity, there may be attendant risks as well as the advantages of simpler management. On the one hand, are planted forests 'sustainable' or are there flaws biologically or even artificially, such as fires arising from arson, which will eventually lead to insuperable problems? On the other hand, are there new directions that may enhance provision of forest products and services from the world's planted forests?

9.8.1 Are planted forests sustainable?

The broad issues of whether using land and devoting resources to planted forest is a sustainable activity from economic, environmental or social dimensions are largely a matter for governments. They relate to industry's and investors' needs, to people's livelihoods and to land use, and fundamentally depend on national policies governing and regulating planted forests and their development, understanding their impacts, and ensuring full stakeholder participation in the process.

Equally pertinent is the narrower focus of the sustainability of the practices surrounding planted forests themselves. Can planted forests be grown indefinitely, rotation after rotation, on the same site without serious risk to their well-being? Specifically, can their long-term productivity be assured?

9.8.2 Do planted forests induce site change?

Two important questions are: (i) do the silvicultural practices commonly applied, such as planting exotic species, use of monocultures, clear felling systems etc., cause site change; and (ii) if so, are such changes more or less favourable to the next crop?

This is a much-researched topic and the overall conclusion is that with good husbandry, such as attending to weed control and correction of known deficiencies, conservation of organic matter, and avoiding damaging harvesting practices, even intensive management of planted forest will not, with very few exceptions, lead to site degradation and unsustainability.

9.8.3 Evidence of productivity change

Data from successive rotations of planted forest

For planted forests, hard evidence of productivity change over successive rotations remains limited, with only a few reliable data. However, what data are available of yields over two, three or four rotations, notably from Chile, South Australia, New Zealand, Swaziland and the south-eastern USA suggest maintenance or improvement in yields over time. In the few instances where yield decline has occurred, it can normally be explained by poor silviculture or poor choice of species rather than any problem with planted forests as such.

Are yields generally increasing?

There is evidence from several countries that productivity of planted forests is gradually increasing over time. More recent plantings are more productive than older ones on any particular site and as they grow stands tend to grow a little better than expected – within-rotation Yield Class/Site quality drift (Evans, 1999). The reasons are thought to be associated with genetic improvement, abandonment of harmful practices such as litter-raking, and a strong fertilization effect from pollutants such as NO_x and increasing atmospheric concentrations of CO_2 (Kahle *et al.*, 2008).

9.9 Risks to Planted Forests and Questions of Ecological Stability

9.9.1 Pest and disease incidence in monocultures

A serious threat to planted forests can arise from a massive build-up of a pest or disease. It is much disputed whether monocultures are more susceptible to devastation from these causes. The broadly accepted ecological principle of stability is that the stability of a community and its constituent species is positively related to its diversity. Following this reasoning, it is argued that, compared with native forest, even-aged monocultures may possess fewer natural constraints on local tree pests and pathogens and thus increase risk of attack. Some evidence supports this but, on the whole, increased susceptibility mostly arises from conditions in planted forests rather than because only one tree species is present.

There are many examples of devastating pest and disease outbreaks, but none has suggested that planted forests are inherently unsustainable, provided that sound research and appropriate investment in control measures are made. As with any cultivated crop, attention must be paid to its health and protection.

9.9.2 Will threats to planted forest increase in the future?

Environmental change

Many of the changes now observed in the composition of the Earth's atmosphere may threaten trees and forests, including planted forests. Rising greenhouse gas concentrations leading to climate change, and increasing atmospheric pollutants of particulates, nitrogen compounds and ozone will stress established planted forests and are likely to increase insect pest risk and disease problems (Evans *et al.*, 2002; Lonsdale and Gibbs, 2002; FAO, 2009). The increasing incidence, scale and impact of extreme weather events (floods, droughts, hurricanes etc.) will need factoring into risk assessments for planted forests investments in the future.

New pests and diseases will emerge

Pests and diseases new to science, new to a country or region, or new as a threat will be facilitated by increasing trade in plant and forest products, illegal introductions arising from international travel, and for exotic species the passing of the 'honeymoon period' of relative freedom from pest and diseases which commonly occurs when they are first introduced.

Droughts, storms and fires

It is difficult to disentangle site effects from those of growing planted forests. Inevitably, large areas of relatively uniform forests will increase the risk of mismatching species and site, and their uniformity may increase risk from storms and possibly fires. Good management, e.g. implementation of FAO's (2006b) Good Practice Guidelines, will minimize these threats.

9.9.3 Enhancing production and other benefits from planted forests

The steady transition from exploitation of native forest to dependence on planted forests is following the path of agriculture. Many of the same biological means to enhance yield are available. They can be expected to increase productivity and alleviate threats, provided information about best practice is known and implemented.

Similarly, beneficial interventions at the local level in a stand of trees, a whole compartment or entire forest estate at the landscape level can aid delivery of non-market benefits, such as biodiversity and better regulated watershed hydrology. Such interventions rarely cause significant loss of production, but meet the legitimate demands of the many different stakeholder interests.

9.9.4 Biological threats from planted forest

Where planting forest involves exotic species, there is a risk that they become invasive and threaten native ecosystems. There are several such examples and steps to contain the risk may be needed, namely, monitoring and investment in research. In addition, centralizing operations, such as in a nursery, may increase the risk of spreading infected germplasm.

9.10 Concluding Comment

Planted forests are becoming the dominant resource for wood production and are playing an increasing role in delivery of some environmental and social benefits. The perception that this is so may be of the most significance as a way of focusing on the true causes of deforestation. This point is important, if perhaps neglected. While most people wrongly attribute the scourge of deforestation simply to excessive logging of natural forests (rather than the real causes of agricultural expansion, urbanisation, settlements etc), a realisation that the great majority of forest products come from just 7% of the world's forest will come as a surprise. Even more surprising is that the bulk of such forests has been planted or has a significant planted component. Appreciating this will help focus attention and political agendas on the real causes of deforestation at a time when native forest resources are more crucial than ever for reasons of biodiversity, carbon storage, climate regulation and soil protection, let alone the needs of the many people groups still dependent on the forests themselves.

There is, perhaps, another surprise. Planted forests are able to fulfil many purposes, not just a single one of production, or of protection, or of amenity. While rarely matching the diversity or environmental benefit of undisturbed native forest, planted forests, as Chapter 6 draws out, can often serve multiple roles at one and the same time.

Planted forests are not a panacea but are a sustainable way of meeting the world's timber requirement from less than 7% of the world's forest area or a mere 2% of land. Thus they can contribute in a singular way to the needs of humanity. This 'domestication' of production did, of course, occur much earlier in agriculture, but just as agriculture has seen astonishing increases in yields through genetic improvement and better husbandry, the same prospect is offered by a planted forest resource. The great advantage for forestry is that this relieves the bulk of the world's forests to serve other purposes, most notably environmental and social. Planted forests have come of age.

Appendix
Planted Forest Areas by Country

A. Del Lungo

Table A.1. Plantation forest area: productive and protective

	Productive							Protective						
	Area ('000 ha)			Annual change rate				Area ('000 ha)			Annual change rate			
	1990	2000	2005	1990–2000 '000 ha/ year	%	2000–2005 '000 ha/ year	%	1990	2000	2005	1990–2000 '000 ha/ year	%	2000–2005 '000 ha/ year	%
Africa														
Eastern and Southern Africa														
Country/area														
Angola	140	134	131	–1	–0.4	–1	–0.4	0	0	0	–	–	–	–
Botswana	–	–	–	–	–	–	–	0	0	0	–	–	–	–
British Indian Ocean Territory	–	–	–	–	–	–	–	0	0	0	–	–	–	–
Comoros	2	2	1	0	0.0	0	–6.2	0	0	0	–	–	–	–
Kenya	238	212	202	–3	–1.2	–2	–1.0	0	0	0	–	–	–	–
Lesotho	4	6	7	0	3.8	0	2.9	0	0	0	–	–	–	–
Madagascar	234	234	234	0	0.0	0	0.0	59	59	59	0	0.0	0	0.0
Malawi	132	180	204	5	3.2	5	2.5	0	0	0	–	–	–	–
Mauritius	12	11	11	0	–0.9	0	0.0	4	4	4	0	0.0	0	0.0
Mayotte	0	0	0	0	0.0	0	0.0	0	0	0	–	–	–	–
Mozambique	38	38	38	0	0.0	0	0.0	0	0	0	–	–	–	–
Namibia	–	–	–	–	–	–	–	0	0	0	–	–	–	–
Réunion	3	3	3	0	0.0	0	–0.8	3	3	3	0	0.0	0	–0.8
Seychelles	5	5	5	0	0.0	0	0.0	0	0	0	–	–	–	–
South Africa	1,204	1,352	1,426	15	1.2	15	1.1	0	0	0	–	–	–	–
Swaziland	135	121	114	–1	–1.1	–1	–1.2	0	0	0	–	–	–	–
Uganda	33	35	36	0	0.6	0	0.6	0	0	0	–	–	–	–
United Republic of Tanzania	150	150	150	0	0.0	0	0.0	0	0	0	–	–	–	–
Zambia	60	75	75	2	2.3	0	0.0	0	0	0	–	–	–	–

–, no data available
0, very small area

Zimbabwe	154	154	154	0	0.0	0	0.0	0	0	0	–	–	–	–
Total (20 countries)	**2,544**	**2,712**	**2,792**	**17**	**0.6**	**16**	**0.6**	**66**	**66**	**66**	**0**	**0.0**	**0**	**0.0**
Northern Africa														
Country/area														
Algeria	6	8	12	0	2.9	1	8.4	614	644	742	3	0.5	20	2.9
Burkina Faso	33	58	71	3	5.8	3	4.1	0	5	5	1	–	0	0.0
Chad	–	–	–	–	–	–	–	11	14	15	0	2.5	0	2.1
Djibouti	–	–	–	–	–	–	–	0	0	0	–	–	–	–
Egypt	–	0	1	0	–	0	26.6	44	59	66	1	2.9	1	2.3
Eritrea	5	11	14	1	8.2	1	4.8	5	11	14	1	8.2	1	4.8
Ethiopia	491	491	491	0	0.0	0	0.0	0	0	0	–	–	–	–
Libyan Arab Jamahiriya	–	–	–	–	–	–	–	217	217	217	0	0.0	0	0.0
Mali	–	60	–	6	–	–	–	0	0	0	–	–	–	–
Mauritania	–	–	–	–	–	–	–	0	10	0	1	–	–	–
Morocco	478	523	563	5	0.9	8	1.5	0	0	0	–	–	–	–
Niger	–	–	–	–	–	–	–	0	73	110	7	–	8	8.7
Somalia	3	3	3	0	0.0	0	0.0	0	0	0	–	–	–	–
Sudan	5,347	4,934	4,728	–41	–0.8	–41	–0.8	764	705	675	–6	–0.8	–6	–0.8
Tunisia	41	129	150	9	12.1	4	3.1	185	294	348	11	4.7	11	3.4
Western Sahara	–	–	–	–	–	–	–	0	0	0	–	–	–	–
Total (16 countries)	**6,404**	**6,218**	**6,033**	**–19**	**–0.3**	**–37**	**–0.6**	**1,840**	**2,031**	**2,192**	**19**	**1.0**	**32**	**1.5**
Western and Central Africa														
Country/area														
Benin	98	109	114	1	1.1	1	0.9	0	0	0	–	–	–	–

(*continued*)

Table A.1. Continued

	Productive							Protective						
	Area ('000 ha)			Annual change rate				Area ('000 ha)			Annual change rate			
				1990–2000		2000–2005					1990–2000		2000–2005	
	1990	2000	2005	'000 ha/ year	%	'000 ha/ year	%	1990	2000	2005	'000 ha/ year	%	'000 ha/ year	%
Burundi	–	86	86	9	–	0	0.0	0	0	0	–	–	–	–
Cameroon	10	9	8	0	–0.9	0	–2.3	78	71	63	–1	–0.9	–2	–2.3
Cape Verde	46	66	67	2	3.6	0	0.4	12	16	17	0	3.6	0	0.4
Central African Republic	2	4	5	0	8.9	0	4.6	0	0	0	–	–	–	–
Congo	51	51	51	0	0.0	0	0.0	0	0	0	–	–	–	–
Côte d'Ivoire	154	261	337	11	5.4	15	5.2	0	0	0	–	–	–	–
Democratic Republic of the Congo	70	68	67	0	–0.4	0	–0.2	30	29	29	0	–0.4	0	–0.2
Equatorial Guinea	–	–	–	–	–	–	–	0	0	0	–	–	–	–
Gabon	36	36	36	0	0.0	0	0.0	0	0	0	–	–	–	–
Gambia	0	0	0	0	0.0	0	0.0	0	0	0	–	–	–	–
Ghana	50	60	160	1	1.8	20	21.7	0	0	0	–	–	–	–
Guinea	15	20	30	1	2.9	2	8.4	2	2	3	0	2.1	0	1.6
Guinea-Bissau	0	0	1	0	5.8	0	7.4	0	0	0	0	7.2	0	14.9
Liberia	8	8	8	0	0.0	0	0.0	0	0	0	–	–	–	–
Nigeria	251	316	349	7	2.3	7	2.0	0	0	0	–	–	–	–
Rwanda	217	248	367	3	1.3	24	8.2	31	35	52	0	1.2	3	8.2
Saint Helena	–	–	–	–	–	–	–	0	0	0	–	–	–	–
São Tomé and Príncipe	–	–	–	–	–	–	–	0	0	0	–	–	–	–
Senegal	185	280	332	10	4.2	10	3.5	20	26	33	1	2.7	1	4.9
Sierra Leone	2	3	3	0	2.7	0	2.9	0	0	0	–	–	–	–
Togo	19	27	30	1	3.6	1	2.1	5	7	8	0	3.4	0	2.7

Total (22 countries)	**1,215**	**1,651**	**2,051**	**44**	**3.1**	**80**	**4.4**	**177**	**187**	**204**	**1**	**0.5**	**3**	**1.8**
Total Africa (58 Countries)	**10,163**	**10,581**	**10,876**	**42**	**0.4**	**59**	**0.6**	**2,083**	**2,283**	**2,462**	**20**	**0.9**	**36**	**1.5**
Asia														
East Asia														
Country/area														
China	17,131	21,765	28,530	463	2.4	1,353	5.6	1,335	2,159	2,839	82	4.9	136	5.6
Democratic People's Republic of Korea	–	–	–	–	–	–	–	0	0	0	–	–	–	–
Japan	–	–	–	–	–	–	–	10,287	10,331	10,321	4	0.0	–2	0.0
Mongolia	30	75	112	5	9.6	7	8.4	0	0	0	–	–	–	–
Republic of Korea	748	1,188	1,364	44	4.7	35	2.8	0	0	0	–	–	–	–
Total (5 countries)	**17,909**	**23,028**	**30,006**	**512**	**2.5**	**1,396**	**5.4**	**11,622**	**12,490**	**13,160**	**87**	**0.7**	**134**	**1.1**
South and South-east Asia														
Country/area														
Bangladesh	173	195	195	2	1.2	0	0.0	66	81	84	2	2.1	1	0.7
Bhutan	1	1	2	0	0.0	0	14.9	0	0	0	–	–	–	–
Brunei Darussalam	–	–	–	–	–	–	–	0	0	0	–	–	–	–
Cambodia	67	72	59	1	0.7	–3	–3.9	0	0	0	–	–	–	–
India	637	915	1,053	28	3.7	28	2.8	1,317	1,890	2,173	57	3.7	57	2.8
Indonesia	2,209	3,002	3,399	79	3.1	79	2.5	0	0	0	–	–	–	–
Lao People's Democratic Republic	3	98	223	10	41.7	25	17.9	1	1	1	0	0.0	0	0.0
Malaysia	1,956	1,659	1,573	–30	–1.6	–17	–1.1	0	0	0	–	–	–	–
Maldives	–	–	–	–	–	–	–	0	0	0	–	–	–	–

(continued)

Table A.1. Continued

	Productive							Protective						
	Area ('000 ha)			Annual change rate				Area ('000 ha)			Annual change rate			
				1990–2000		2000–2005					1990–2000		2000–2005	
	1990	2000	2005	'000 ha/ year	%	'000 ha/ year	%	1990	2000	2005	'000 ha/ year	%	'000 ha/ year	%
Myanmar	323	571	696	25	5.9	25	4.1	71	125	153	5	5.8	6	4.1
Nepal	40	42	43	0	0.5	0	0.5	9	10	10	0	1.1	0	0.0
Pakistan	234	296	318	6	2.4	4	1.4	0	0	0	–	–	–	–
Philippines	389	321	304	–7	–1.9	–3	–1.1	1,391	531	316	–86	–9.2	–43	–9.9
Singapore	–	–	–	–	–	–	–	0	0	0	–	–	–	–
Sri Lanka	221	198	171	–2	–1.1	–5	–2.8	21	23	24	0	0.9	0	0.7
Thailand	1,979	1,996	1,997	2	0.1	0	0.0	661	1,081	1,102	42	5.0	4	0.4
Timor–Leste	–	–	–	–	–	–	–	29	43	43	1	4.0	0	0.0
Viet Nam	664	1,384	1,792	72	7.6	82	5.3	303	666	903	36	8.2	47	6.3
Total (18 countries)	**8,896**	**10,750**	**11,825**	**185**	**1.9**	**215**	**1.9**	**3,869**	**4,451**	**4,809**	**58**	**1.4**	**71**	**1.6**
Western and Central Asia														
Country/area														
Armenia	–	–	–	–	–	–	–	14	11	10	0	–2.4	0	–1.9
Azerbaijan	–	–	–	–	–	–	–	20	20	20	0	0.0	0	0.0
Bahrain	–	–	–	–	–	–	–	0	0	0	0	5.3	0	4.0
Cyprus	–	–	–	–	–	–	–	3	3	5	0	0.0	0	10.8
Georgia	–	–	–	–	–	–	–	54	60	61	1	1.1	0	0.2
Iran (Islamic Republic of)	616	616	616	0	0.0	0	0.0	0	0	0	–	–	–	–
Iraq	–	–	–	–	–	–	–	15	15	13	0	0.0	0	–2.8
Israel	–	–	–	–	–	–	–	84	94	101	1	1.1	1	1.4
Jordan	–	–	–	–	–	–	–	40	40	40	0	0.0	0	0.0

Kazakhstan	–	–	–	–	–	–	–	1,034	1,056	909	2	0.2	–29	–3.0
Kuwait	–	–	–	–	–	–	–	3	5	6	0	3.5	0	2.7
Kyrgyzstan	19	22	24	0	1.5	0	1.4	26	37	42	1	3.5	1	2.7
Lebanon	–	–	–	–	–	–	–	0	0	0	–	–	–	–
Oman	–	–	–	–	–	–	–	2	2	2	0	0.0	0	0.0
Palestine, Occupied Territories	–	–	–	–	–	–	–	0	0	0	–	–	–	–
Qatar	–	–	–	–	–	–	–	0	0	0	–	–	–	–
Saudi Arabia	–	–	–	–	–	–	–	0	0	0	–	–	–	–
Syrian Arab Republic	–	–	–	–	–	–	–	175	234	264	6	3.0	6	2.4
Tajikistan	22	22	22	0	0.0	0	0.0	54	44	44	–1	–2.0	0	0.0
Turkey	1,459	1,763	1,916	30	1.9	31	1.7	380	541	621	16	3.6	16	2.8
Turkmenistan	–	–	–	–	–	–	–	0	0	0	–	–	–	–
United Arab Emirates	–	–	–	–	–	–	–	245	310	312	7	2.4	0	0.1
Uzbekistan	4	5	5	0	2.3	0	0.0	26	46	56	2	5.9	2	4.0
Yemen	–	–	–	–	–	–	–	0	0	0	–	–	–	–
Total (24 countries)	**2,120**	**2,428**	**2,583**	**31**	**1.4**	**31**	**1.2**	**2,175**	**2,518**	**2,505**	**34**	**1.5**	**–3**	**–0.1**
Total Asia (47 countries)	**28,925**	**36,206**	**44,414**	**728**	**2.3**	**1,642**	**4.2**	**17,666**	**19,459**	**20,474**	**179**	**1.0**	**203**	**1.0**
Oceania														
Oceania														
Country/area														
American Samoa	–	–	–	–	–	–	–	0	0	0	–	–	–	–
Australia	1,023	1,485	1,766	46	3.8	56	3.5	0	0	0	–	–	–	–
Cook Islands	–	–	–	–	–	–	–	1	1	1	0	8.2	0	0.0
Fiji	80	101	101	2	2.4	0	0.0	0	0	0	–	–	–	–
French Polynesia	10	10	10	0	0.0	0	0.0	0	0	0	–	–	–	–

(continued)

Table A.1. Continued

	Productive							Protective						
	Area ('000 ha)			Annual change rate				Area ('000 ha)			Annual change rate			
				1990–2000		2000–2005					1990–2000		2000–2005	
	1990	2000	2005	'000 ha/ year	%	'000 ha/ year	%	1990	2000	2005	'000 ha/ year	%	'000 ha/ year	%
Guam	–	–	–	–	–	–	–	0	0	0	–	–	–	–
Kiribati	–	–	–	–	–	–	–	0	0	0	–	–	–	–
Marshall Islands	–	–	–	–	–	–	–	0	0	0	–	–	–	–
Micronesia (Federated States of)	–	–	–	–	–	–	–	0	0	0	–	–	–	–
Nauru	–	–	–	–	–	–	–	0	0	0	–	–	–	–
New Caledonia	10	10	10	0	0.0	0	0.0	0	0	0	–	–	–	–
New Zealand	1,261	1,767	1,832	51	3.4	13	0.7	0	2	20	0	–	4	58.5
Niue	0	0	0	0	40.0	0	0.0	0	0	0	–	–	–	–
Northern Mariana Islands	–	–	–	–	–	–	–	0	0	0	–	–	–	–
Palau	–	–	–	–	–	–	–	0	0	0	–	–	–	–
Papua New Guinea	63	82	92	2	2.8	2	2.3	0	0	0	–	–	–	–
Pitcairn	–	–	–	–	–	–	–	0	0	0	–	–	–	–
Samoa	–	21	21	2	–	0	0.0	0	11	11	1	–	0	0.0
Solomon Islands	–	–	–	–	–	–	–	0	0	0	–	–	–	–
Tokelau	–	–	–	–	–	–	–	0	0	0	–	–	–	–
Tonga	0	0	0	0	0.0	0	0.0	0	0	0	–	–	–	–
Tuvalu	–	–	–	–	–	–	–	0	0	0	–	–	–	–
Vanuatu	–	–	–	–	–	–	–	0	0	0	–	–	–	–
Wallis and Futuna Islands	0	1	1	0	9.6	0	1.2	0	0	0	–	–	–	–

Total Oceania (24 countries)	**2,447**	**3,477**	**3,833**	**103**	**3.6**	**71**	**2.0**	**1**	**14**	**32**	**1**	**39.4**	**4**	**18.1**
Europe														
Europe														
Country/area														
Albania	11	5	3	–1	–8.0	0	–11.2	93	92	86	0	–0.1	–1	–1.4
Andorra	–	–	–	–	–	–	–	0	0	0	–	–	–	–
Austria	988	1,003	–	2	0.2	–	–	0	0	0	–	–	–	–
Belarus	2	2	2	0	1.6	0	0.0	0	0	0	–	–	–	–
Belgium	303	284	275	–2	–0.7	–2	–0.6	0	0	0	–	–	–	–
Bosnia and Herzegovina	–	142	142	14	–	0	0.0	0	0	0	–	–	–	–
Bulgaria	22	31	35	1	3.4	1	2.7	19	17	16	0	–0.9	0	–1.0
Channel Islands	–	–	–	–	–	–	–	0	0	0	–	–	–	–
Croatia	56	60	61	0	0.7	0	0.3	0	0	0	–	–	–	–
Czech Republic	–	–	–	–	–	–	–	0	0	0	–	–	–	–
Denmark	257	271	281	1	0.5	2	0.7	34	34	34	0	0.0	0	0.0
Estonia	–	1	1	0	–	0	0.0	0	0	0	–	–	–	–
Faeroe Islands	–	–	–	–	–	–	–	0	0	0	–	–	–	–
Finland	–	–	–	–	–	–	–	0	0	0	–	–	–	–
France	1,842	1,936	1,968	9	0.5	6	0.3	0	0	0	–	–	–	–
Germany	–	–	–	–	–	–	–	0	0	0	–	–	–	–
Gibraltar	–	–	–	–	–	–	–	0	0	0	–	–	–	–
Greece	–	–	–	–	–	–	–	118	129	134	1	0.9	1	0.8
Holy See	–	–	–	–	–	–	–	0	0	0	–	–	–	–
Hungary	39	439	455	40	27.3	3	0.7	392	89	91	–30	–13.8	0	0.5
Iceland	4	12	17	1	11.4	1	7.6	4	10	12	1	10.2	0	4.1
Ireland	350	519	579	17	4.0	12	2.2	0	0	0	–	–	–	–
Isle of Man	–	–	–	–	–	–	–	0	0	0	–	–	–	–
Italy	289	144	146	–15	–6.7	0	0.3	0	0	0	–	–	–	–

(continued)

Table A.1. Continued

	Productive							Protective						
	Area ('000 ha)			Annual change rate				Area ('000 ha)			Annual change rate			
	1990	2000	2005	1990–2000 '000 ha/ year	%	2000–2005 '000 ha/ year	%	1990	2000	2005	1990–2000 '000 ha/ year	%	2000–2005 '000 ha/ year	%
Latvia	–	0	1	0	–	0	90.1	0	0	0	–	–	–	–
Liechtenstein	0	0	0	0	4.1	0	0.0	0	0	0	–	–	–	–
Lithuania	84	95	100	1	1.2	1	1.0	40	42	41	0	0.5	0	–0.5
Luxembourg	28	28	28	0	0.0	0	0.0	0	0	0	–	–	–	–
Malta	–	–	–	–	–	–	–	0	0	0	0	0.0	0	0.0
Monaco	–	–	–	–	–	–	–	0	0	0	–	–	–	–
Netherlands	4	4	4	0	0.0	0	0.0	0	0	0	–	–	–	–
Norway	222	255	262	3	1.4	1	0.5	0	0	0	–	–	–	–
Poland	32	32	32	0	0.0	0	0.0	0	0	0	–	–	–	–
Portugal	383	867	1,067	48	8.5	40	4.2	167	167	167	0	0.0	0	0.0
Republic of Moldova	1	1	1	0	0.0	0	0.0	0	0	0	–	–	–	–
Romania	92	92	92	0	0.0	0	0.0	57	57	57	0	0.0	0	0.0
Russian Federation	9,244	10,712	11,888	147	1.5	235	2.1	3,407	4,648	5,075	124	3.2	85	1.8
San Marino	–	–	–	–	–	–	–	0	0	0	–	–	–	–
Serbia and Montenegro	39	39	39	0	0.0	0	0.0	0	0	0	–	–	–	–
Slovakia	21	18	17	0	–1.5	0	–1.1	2	2	2	0	0.0	0	0.0
Slovenia	–	–	–	–	–	–	–	0	0	0	–	–	–	–
Spain	1,126	1,356	1,471	23	1.9	23	1.6	0	0	0	–	–	–	–
Sweden	523	619	667	10	1.7	10	1.5	0	0	0	–	–	–	–

Switzerland	3	4	4	0	2.9	0	0.0	0	0	0	–	–	–	–
The former Yugoslav Republic of Macedonia	30	30	30	0	0.0	0	0.0	0	0	0	–	–	–	–
Ukraine	84	82	81	0	–0.2	0	–0.2	241	285	307	4	1.7	4	1.5
United Kingdom	1,862	1,914	1,902	5	0.3	–2	–0.1	15	20	22	1	2.9	0	1.9
Total Europe (47 countries)	**17,942**	**20,997**	**21,651**	**306**	**1.6**	**131**	**0.6**	**4,588**	**5,591**	**6,044**	**100**	**2.0**	**90**	**1.6**
North and Central America														
North America														
Country/area														
Canada	–	–	–	–	–	–	–	0	0	0	–	–	–	–
Greenland	–	–	–	–	–	–	–	0	0	0	–	–	–	–
Mexico	–	11	72	1	–	12	45.6	0	1,047	986	105	–	–12	–1.2
Saint Pierre and Miquelon	–	–	–	–	–	–	–	0	0	0	–	–	–	–
United States of America	10,305	16,274	17,061	597	4.7	157	0.9	0	0	0	–	–	–	–
Total (5 countries)	**10,305**	**16,285**	**17,133**	**598**	**4.7**	**170**	**1.0**	**0**	**1,047**	**986**	**105**	**–**	**–12**	**–1.2**
Central America														
Country/area														
Belize	–	–	–	–	–	–	–	0	0	0	–	–	–	–
Costa Rica	–	1	1	0	–	0	0.0	0	2	3	0	–	0	8.4
El Salvador	6	6	6	0	0.0	0	0.0	0	0	0	–	–	–	–
Guatemala	32	88	122	6	10.6	7	6.8	0	0	0	–	–	–	–
Honduras	–	–	–	–	–	–	–	31	26	30	–1	–1.7	1	2.9
Nicaragua	4	46	51	4	27.7	1	2.1	0	0	0	–	–	–	–
Panama	9	42	60	3	16.5	4	7.7	1	1	1	0	–6.7	0	14.9
Total (7 countries)	**51**	**183**	**240**	**13**	**13.6**	**12**	**5.6**	**32**	**29**	**34**	**0**	**–1.2**	**1**	**3.6**

(continued)

Table A.1. Continued

	Productive							Protective						
	Area ('000 ha)			Annual change rate				Area ('000 ha)			Annual change rate			
				1990–2000		2000–2005					1990–2000		2000–2005	
	1990	2000	2005	'000 ha/ year	%	'000 ha/ year	%	1990	2000	2005	'000 ha/ year	%	'000 ha/ year	%
Caribbean														
Country/area														
Anguilla	–	–	–	–	–	–	–	0	0	0	–	–	–	–
Antigua and Barbuda	–	–	–	–	–	–	–	0	0	0	–	–	–	–
Aruba	–	–	–	–	–	–	–	0	0	0	–	–	–	–
Bahamas	–	–	–	–	–	–	–	0	0	0	–	–	–	–
Barbados	–	–	–	–	–	–	–	0	0	0	–	–	–	–
Bermuda	–	–	–	–	–	–	–	0	0	0	–	–	–	–
British Virgin Islands	–	–	–	–	–	–	–	0	0	0	–	–	–	–
Cayman Islands	–	–	–	–	–	–	–	0	0	0	–	–	–	–
Cuba	197	197	230	0	–0.1	7	3.1	149	145	164	0	–0.3	4	2.5
Dominica	–	0	0	0	–	0	0.0	0	0	0	–	–	–	–
Dominican Republic	–	–	–	–	–	–	–	0	0	0	–	–	–	–
Grenada	0	0	0	0	0.0	0	0.0	0	0	0	–	–	–	–
Guadeloupe	3	1	1	0	–10.4	0	0.0	0	0	0	–	–	–	–
Haiti	12	20	24	1	5.2	1	3.7	0	0	0	–	–	–	–
Jamaica	9	8	8	0	–0.8	0	0.0	6	6	6	0	0.5	0	0.3
Martinique	1	1	1	0	0.0	0	0.0	0	0	0	–	–	–	–
Montserrat	–	–	–	–	–	–	–	0	0	0	–	–	–	–

Netherlands Antilles	–	–	–	–	–	–	–	0	0	0	–	–	–	–
Puerto Rico	–	–	–	–	–	–	–	0	0	0	–	–	–	–
Saint Kitts and Nevis	–	–	–	–	–	–	–	0	0	0	–	–	–	–
Saint Lucia	–	–	–	–	–	–	–	0	0	0	–	–	–	–
Saint Vincent and the Grenadines	0	0	0	0	7.6	0	5.1	0	0	0	–	–	–	–
Trinidad and Tobago	15	15	15	0	0.0	0	0.0	0	0	0	–	–	–	–
Turks and Caicos Islands	–	–	–	–	–	–	–	0	0	0	–	–	–	–
United States Virgin Islands	–	–	–	–	–	–	–	0	0	0	–	–	–	–
Total (25 countries)	**239**	**243**	**280**	**0**	**0.2**	**7**	**2.9**	**155**	**151**	**170**	**0**	**–0.2**	**4**	**2.4**
Total North and Central America (37 countries)	**10,595**	**16,711**	**17,653**	**612**	**4.7**	**189**	**1.1**	**187**	**1,227**	**1,190**	**104**	**20.7**	**–7**	**–0.6**
South America														
South America														
Country/area														
Argentina	769	1,078	1,229	31	3.4	30	2.7	0	0	0	–	–	–	–
Bolivia	20	20	20	0	0.0	0	0.0	0	0	0	–	–	–	–
Brazil	5,070	5,279	5,384	21	0.4	21	0.4	0	0	0	–	–	–	–
Chile	1,741	2,354	2,661	61	3.1	61	2.5	0	0	0	–	–	–	–
Colombia	130	241	312	11	6.4	14	5.3	7	13	16	1	7.2	1	4.2
Ecuador	–	162	164	16	–	1	0.3	0	0	0	–	–	–	–
Falkland Islands	–	–	–	–	–	–	–	0	0	0	–	–	–	–
French Guiana	1	1	1	0	0.0	0	0.0	0	0	0	–	–	–	–
Guyana	–	–	–	–	–	–	–	0	0	0	–	–	–	–
Paraguay	23	36	43	1	4.6	1	3.6	0	0	0	–	–	–	–

(continued)

Table A.1. Continued

	Productive							Productive						
	Area ('000 ha)			Annual change rate				Area ('000 ha)			Annual change rate			
	1990	2000	2005	1990–2000 '000 ha/ year	%	2000–2005 '000 ha/ year	%	1990	2000	2005	1990–2000 '000 ha/ year	%	2000–2005 '000 ha/ year	%
Peru	263	715	754	45	10.5	8	1.1	0	0	0	–	–	–	–
South Georgia and the South Sandwich Islands	–	–	–	–	–	–	–	0	0	0	–	–	–	–
Suriname	7	7	7	0	0.0	0	0.0	0	0	0	–	–	–	–
Uruguay	197	655	751	46	12.8	19	2.8	4	14	15	1	13.3	0	1.4
Venezuela (Bolivarian Republic of)	873	836	806	–4	–0.4	–6	–0.7	28	27	26	0	–0.4	0	–0.8
Total (15 countries)	**9,094**	**11,383**	**12,132**	**229**	**2.3**	**150**	**1.3**	**39**	**54**	**57**	**2**	**3.4**	**1**	**1.1**
WORLD	**79,165**	**99,356**	**110,560**	**2,019**	**2.3**	**2,241**	**2.2**	**24,562**	**28,628**	**30,259**	**407**	**1.5**	**326**	**1.1**

Source: FAO Global Forest Resources Assessment 2005.

Table A.2. Planted semi-natural forests area; productive and protective: 61 sampled countries

	Productive							Protective						
	Area ('000 ha)			Annual change rate				Area ('000 ha)			Annual change rate			
	1990	2000	2005	1990–2000 '000 ha/ year	%	2000–2005 '000 ha/ year	%	1990	2000	2005	1990–2000 '000 ha/ year	%	2000–2005 '000 ha/ year	%
Africa														
Eastern and Southern Africa														
Country/area														
South Africa	–	–	–	–	–	–	–	–	–	–	–	–	–	–
Total (1 country)	**0**	**0**	**0**	**–**	**–**	**–**	**–**	**0**	**0**	**0**	**–**	**–**	**–**	**–**
Northern Africa														
Country/area														
Algeria	5	11	14	1	8.3	1	6.0	234	255	270	2	0.9	3	1.1
Sudan	1,039	993	948	–5	–0.5	–9	–0.9	260	248	267	–1	–0.5	4	1.5
Total (2 countries)	**1,044**	**1,003**	**963**	**–4**	**0.0**	**–8**	**–0.8**	**494**	**504**	**538**	**1**	**0.0**	**7**	**1.3**
Western and Central Africa														
Country/area														
Cameroon	–	–	–	–	–	–	–	–	–	–	–	–	–	–
Democratic Republic of the Congo	–	–	–	–	–	–	–	–	–	–	–	–	–	–
Nigeria	–	–	–	–	–	–	–	–	–	–	–	–	–	–

(*continued*)

Table A.2. Continued

	Productive							Protective						
	Area ('000 ha)			Annual change rate				Area ('000 ha)			Annual change rate			
	1990	2000	2005	1990–2000 '000 ha/ year	%	2000–2005 '000 ha/ year	%	1990	2000	2005	1990–2000 '000 ha/ year	%	2000–2005 '000 ha/ year	%
Total (3 countries)	**0**	**0**	**0**	**–**	**–**	**–**	**–**	**0**	**0**	**0**	**–**	**–**	**–**	**–**
Total Africa (6 countries)	**1,044**	**1,003**	**963**	**–4**	**0.0**	**–8**	**–0.8**	**494**	**504**	**538**	**1**	**0.0**	**7**	**1.3**
Asia														
East Asia														
Country/area														
China	16,944	20,123	25,572	318	1.7	1,090	4.9	8,345	10,366	14,385	202	2.2	804	6.8
Japan	–	–	–	–	–	–	–	–	–	–	–	–	–	–
Total (2 countries)	**16,944**	**20,123**	**25,572**	**318**	**0.0**	**1,090**	**4.9**	**8,345**	**10,366**	**14,385**	**202**	**0.0**	**804**	**6.8**
South and South-east Asia														
Country/area														
India	18,997	17,562	16,081	–143	–0.8	–296	–1.7	9,786	10,764	10,721	98	1.0	–9	–0.1
Indonesia	–	–	–	–	–	–	–	–	–	–	–	–	–	–
Malaysia	–	–	–	–	–	–	–	–	–	–	–	–	–	–
Myanmar	–	–	–	–	–	–	–	–	–	–	–	–	–	–
Pakistan	–	–	–	–	–	–	–	–	–	–	–	–	–	–
Philippines	–	–	–	–	–	–	–	–	–	–	–	–	–	–
Thailand	–	–	–	–	–	–	–	–	–	–	–	–	–	–
Viet Nam	–	–	–	–	–	–	–	–	–	–	–	–	–	–

Total (8 countries)	**18,997**	**17,562**	**16,081**	**–143**	**0.0**	**–296**	**–1.7**	**9,786**	**10,764**	**10,721**	**98**	**0.0**	**–9**	**–0.1**
Western and Central Asia														
Country/area														
Georgia	–	–	–	–	–	–	–	–	–	–	–	–	–	–
Iran	24	50	5	3	7.5	–9	–38.1	–	–	–	–	–	–	–
Turkey	62	87	100	2	3.4	2	2.7	146	203	232	6	3.4	6	2.7
Total (3 countries)	**87**	**137**	**104**	**5**	**0.0**	**–7**	**–5.4**	**146**	**203**	**232**	**6**	**0.0**	**6**	**2.7**
Total Asia (13 countries)	**36,027**	**37,822**	**41,758**	**179**	**0.0**	**787**	**2.0**	**18,277**	**21,333**	**25,338**	**306**	**0.0**	**801**	**3.5**
Europe														
Europe														
Country/area														
Albania	–	–	–	–	–	–	–	–	–	–	–	–	–	–
Austria	271	275	275	0	0.2	0	0.0	–	–	–	–	–	–	–
Belarus	1,330	1,432	1,508	10	0.7	15	1.0	–	–	–	–	–	–	–
Belgium	157	161	149	0	0.3	–2	–1.6	–	–	–	–	–	–	–
Bosnia and Herzegovina	926	758	758	–17	–2.0	0	0.0	–	–	–	–	–	–	–
Bulgaria	526	425	399	–10	–2.1	–5	–1.2	466	460	432	–1	–0.1	–6	–1.2
Croatia	–	–	–	–	–	–	–	–	–	–	–	–	–	–
Czech Republic	2,449	2,405	2,415	–4	–0.2	2	0.1	63	87	88	2	3.3	0	0.1
Denmark	118	135	139	2	1.4	1	0.5	16	17	17	0	0.8	0	–0.1
Estonia	–	462	505	46	–	9	1.8	–	–	–	–	–	–	–
Finland	3,934	4,843	5,270	91	2.1	85	1.7	–	–	–	–	–	–	–
France	119	127	129	1	0.7	0	0.3	8	7	7	0	–1.3	0	0.3
Germany	4,540	4,513	4,441	–3	–0.1	–14	–0.3	1,904	2,133	2,204	23	1.1	14	0.7
Greece	–	–	–	–	–	–	–	–	–	–	–	–	–	–
Hungary	26	442	459	42	32.9	3	0.8	256	89	92	–17	–10.0	1	0.6

(continued)

Table A.2. Continued

	Productive							Protective						
	Area ('000 ha)			Annual change rate				Area ('000 ha)			Annual change rate			
				1990–2000		2000–2005					1990–2000		2000–2005	
	1990	2000	2005	'000 ha/ year	%	'000 ha/ year	%	1990	2000	2005	'000 ha/ year	%	'000 ha/ year	%
Ireland	90	90	90	0	0.0	0	0.0	–	–	–	–	–	–	–
Italy	–	–	–	–	–	–	–	–	–	–	–	–	–	–
Latvia	–	563	567	56	–	1	0.2	–	69	77	7	–	2	2.2
Lithuania	231	257	288	3	1.1	6	2.3	77	86	96	1	1.1	2	2.3
Netherlands	224	222	206	0	–0.1	–3	–1.5	56	56	69	0	–0.1	3	4.3
Norway	1,126	1,319	1,420	19	1.6	20	1.5	–	–	–	–	–	–	–
Poland	5,769	5,595	5,584	–17	–0.3	–2	0.0	2,715	3,013	3,141	30	1.0	26	0.8
Portugal	278	334	345	6	1.9	2	0.6	121	65	54	–6	–6.1	–2	–3.6
Republic of Moldova	–	–	–	–	–	–	–	–	–	–	–	–	–	–
Romania	3,220	3,155	3,155	–7	–0.2	0	0.0	1,996	1,955	1,955	–4	–0.2	0	0.0
Russian Federation	–	–	–	–	–	–	–	–	–	–	–	–	–	–
Slovakia	665	666	668	0	0.0	0	0.0	159	159	160	0	0.0	0	0.0
Slovenia	34	36	37	0	0.6	0	0.5	–	–	–	–	–	–	–
Spain	356	446	494	9	2.3	10	2.1	–	–	–	–	–	–	–
Sweden	7,199	8,234	9,297	104	1.4	212	2.5	–	–	–	–	–	–	–
Switzerland	48	50	50	0	0.3	0	0.1	–	–	–	–	–	–	–
Ukraine	3,018	2,874	2,991	–14	–0.5	23	0.8	1,294	1,514	1,408	22	1.6	–21	–1.4
United Kingdom	–	–	–	–	–	–	–	88	211	264	12	9.1	11	4.6
Total Europe (33 countries)	**36,652**	**39,820**	**41,638**	**317**	**0.0**	**364**	**0.9**	**9,218**	**9,919**	**10,062**	**70**	**0.0**	**29**	**0.3**

North and Central America														
North America														
Country/area														
Canada	3,976	8,147	10,206	417	7.4	412	4.6	–	–	–	–	–	–	–
United States of America	–	–	–	–	–	–	–	–	–	–	–	–	–	–
Total (2 countries)	**3,976**	**8,147**	**10,206**	**417**	**0.0**	**412**	**4.6**	**0**	**0**	**0**	**–**	**–**	**–**	**–**
Total North and Central America (2 countries)	**3,976**	**8,147**	**10,206**	**417**	**0.0**	**412**	**4.6**	**0**	**0**	**0**	**–**	**–**	**–**	**–**
Oceania														
Oceania														
Country/area														
Australia	–	–	–	–	–	–	–	–	–	–	–	–	–	–
New Zealand	–	–	–	–	–	–	–	–	–	–	–	–	–	–
Total Oceania (2 countries)	**0**	**0**	**0**	**–**	**–**	**–**	**–**	**0**	**0**	**0**	**–**	**–**	**–**	**–**
South America														
South America														
Country/area														
Argentina	–	–	–	–	–	–	–	–	–	–	–	–	–	–
Brazil	–	–	–	–	–	–	–	–	–	–	–	–	–	–
Chile	25	25	25	0	0.0	0	0.0	–	–	–	–	–	–	–
Uruguay	–	–	–	–	–	–	–	–	–	–	–	–	–	–
Venezuela	–	–	–	–	–	–	–	–	–	–	–	–	–	–
Total (5 countries)	**25**	**25**	**25**	**0**	**0.0**	**0**	**0.0**	**0**	**0**	**0**	**–**	**–**	**–**	**–**
WORLD	**77,724**	**86,817**	**94,590**	**909**	**0.0**	**1,554**	**1.7**	**27,990**	**31,756**	**35,938**	**377**	**0.0**	**836**	**2.5**

Source: FAO Global Planted Forests Thematic Study – Results and Analysis (FAO, 2006c).

Table A.3. Total planted forest area productive and protective: 61 sampled countries

	Productive							Protective						
	Area ('000 ha)			Annual change rate				Area ('000 ha)			Annual change rate			
	1990	2000	2005	1990–2000 '000 ha/ year	%	2000–2005 '000 ha/ year	%	1990	2000	2005	1990–2000 '000 ha/ year	%	2000–2005 '000 ha/ year	%
Africa														
Eastern and Southern Africa														
Country/area														
South Africa	1,204	1,352	1,426	15	1.2	15	1.1	–	–	–	–	–	–	–
Total (1 country)	**1,204**	**1,352**	**1,426**	**15**	**0.0**	**15**	**1.1**	**0**	**0**	**0**	**–**	**–**	**–**	**–**
Northern Africa														
Country/area														
Algeria	11	19	26	1	5.6	2	7.1	848	899	1,012	5	0.6	23	2.4
Sudan	6,385	5,927	5,677	–46	–0.7	–50	–0.9	1,024	953	943	–7	–0.7	–2	–0.2
Total (2 countries)	**6,396**	**5,946**	**5,703**	**–45**	**0.0**	**–49**	**–0.8**	**1,872**	**1,852**	**1,955**	**–2**	**0.0**	**21**	**1.1**
Western and Central Africa														
Country/area														
Cameroon	10	9	8	0	–0.9	0	–2.3	78	71	63	–1	–0.9	–2	–2.3
Democratic Republic of the Congo	70	68	67	0	–0.4	0	–0.2	30	29	29	0	–0.4	0	–0.2
Nigeria	251	316	349	7	2.3	7	2.0	–	–	–	–	–	–	–

Total (3 countries)	**332**	**393**	**424**	**6**	**0.0**	**6**	**1.5**	**108**	**100**	**92**	**–1**	**0.0**	**–2**	**–1.7**
Total Africa (6 countries)	**7,932**	**7,691**	**7,553**	**–24**	**0.0**	**–28**	**–0.4**	**1,980**	**1,952**	**2,047**	**–3**	**0.0**	**19**	**1.0**
Asia														
East Asia														
Country/area														
China	34,075	41,888	54,102	781	2.1	2,443	5.3	9,680	12,525	17,224	284	2.6	940	6.6
Japan	–	–	–	–	–	–	–	10,287	10,331	10,321	4	0.0	–2	0.0
Total (2 countries)	**34,075**	**41,888**	**54,102**	**781**	**0.0**	**2,443**	**5.3**	**19,967**	**22,856**	**27,545**	**289**	**0.0**	**938**	**3.8**
South and South-east Asia														
Country/area														
India	19,634	18,477	17,134	–116	–0.6	–268	–1.5	11,103	12,654	12,894	155	1.3	48	0.4
Indonesia	2,209	3,002	3,399	79	3.1	79	2.5	–	–	–	–	–	–	–
Malaysia	1,956	1,659	1,573	–30	–1.6	–17	–1.1	–	–	–	–	–	–	–
Myanmar	323	571	696	25	5.9	25	4.1	71	125	153	5	5.8	6	4.1
Pakistan	234	296	318	6	2.4	4	1.4	–	–	–	–	–	–	–
Philippines	389	321	304	–7	–1.9	–3	–1.1	1,391	531	316	–86	–9.2	–43	–9.9
Thailand	1,979	1,996	1,997	2	0.1	0	0.0	661	1,081	1,102	42	5.0	4	0.4
Viet Nam	664	1,384	1,792	72	7.6	82	5.3	303	666	903	36	8.2	47	6.3
Total (8 countries)	**27,388**	**27,705**	**27,214**	**32**	**0.0**	**–98**	**–0.4**	**13,529**	**15,057**	**15,368**	**153**	**0.0**	**62**	**0.4**
Western and Central Asia														
Country/area														
Georgia	–	–	–	–	–	–	–	54	60	61	1	1.1	0	0.2
Iran (Islamic Republic of)	640	666	621	3	0.4	–9	–1.4	–	–	–	–	–	–	–

(*continued*)

Table A.3. Continued

	Productive							Protective						
	Area ('000 ha)			Annual change rate				Area ('000 ha)			Annual change rate			
	1990	2000	2005	1990–2000 '000 ha/year	%	2000–2005 '000 ha/year	%	1990	2000	2005	1990–2000 '000 ha/year	%	2000–2005 '000 ha/year	%
Turkey	1,521	1,850	2,016	33	2.0	33	1.7	526	744	853	22	3.5	22	2.8
Total (3 countries)	**2,162**	**2,516**	**2,636**	**35**	**0.0**	**24**	**0.9**	**580**	**804**	**914**	**22**	**0.0**	**22**	**2.6**
Total Asia (13 countries)	**63,624**	**72,110**	**83,952**	**849**	**0.0**	**2,369**	**3.1**	**34,076**	**38,718**	**43,826**	**464**	**0.0**	**1,022**	**2.5**
Europe														
Europe														
Country/area														
Albania	11	5	3	–1	–8.0	0	–11.2	93	92	86	0	–0.1	–1	–1.4
Austria	1,259	1,278	1,278	2	0.2	0	0.0	–	–	–	–	–	–	–
Belarus	1,331	1,434	1,510	10	0.7	15	1.0	–	–	–	–	–	–	–
Belgium	460	445	424	–2	–0.3	–4	–1.0	–	–	–	–	–	–	–
Bosnia and Herzegovina	926	900	900	–3	–0.3	0	0.0	–	–	–	–	–	–	–
Bulgaria	548	456	434	–9	–1.8	–4	–1.0	485	477	449	–1	–0.2	–6	–1.2
Croatia	56	60	61	0	0.7	0	0.3	–	–	–	–	–	–	–
Czech Republic	2,449	2,405	2,415	–4	–0.2	2	0.1	63	87	88	2	3.3	0	0.1
Denmark	375	406	420	3	0.8	3	0.7	50	51	51	0	0.2	0	0.0
Estonia	–	463	506	46	–	9	1.8	–	–	–	–	–	–	–
Finland	3,934	4,843	5,270	91	2.1	85	1.7	–	–	–	–	–	–	–
France	1,961	2,063	2,097	10	0.5	7	0.3	8	7	7	0	–1.3	0	0.3
Germany	4,540	4,513	4,441	–3	–0.1	–14	–0.3	1,904	2,133	2,204	23	1.1	14	0.7
Greece	–	–	–	–	–	–	–	118	129	134	1	0.9	1	0.8
Hungary	65	881	914	82	29.8	6	0.7	648	178	183	–47	–12.1	1	0.6

Ireland	440	609	669	17	3.3	12	1.9	–	–	–	–	–	–	–
Italy	289	144	146	–15	–6.7	0	0.3	–	–	–	–	–	–	–
Latvia	–	563	569	56	–	1	0.2	–	69	77	7	–	2	2.2
Lithuania	315	352	388	4	1.1	7	2.0	117	128	137	1	0.9	2	1.4
Netherlands	228	226	210	0	–0.1	–3	–1.5	56	56	69	0	–0.1	3	4.3
Norway	1,348	1,574	1,682	23	1.6	22	1.3	–	–	–	–	–	–	–
Poland	5,801	5,627	5,616	–17	–0.3	–2	0.0	2,715	3,013	3,141	30	1.0	26	0.8
Portugal	661	1,201	1,412	54	6.2	42	3.3	288	232	221	–6	–2.2	–2	–0.9
Republic of Moldova	1	1	1	0	0.0	0	0.0	–	–	–	–	–	–	–
Romania	3,312	3,247	3,247	–7	–0.2	0	0.0	2,053	2,012	2,012	–4	–0.2	0	0.0
Russian Federation	9,244	10,712	11,888	147	1.5	235	2.1	3,407	4,648	5,075	124	3.2	85	1.8
Slovakia	686	684	685	0	0.0	0	0.0	161	161	162	0	0.0	0	0.0
Slovenia	34	36	37	0	0.6	0	0.5	–	–	–	–	–	–	–
Spain	1,482	1,802	1,965	32	2.0	33	1.7	–	–	–	–	–	–	–
Sweden	7,722	8,853	9,964	113	1.4	222	2.4	–	–	–	–	–	–	–
Switzerland	51	54	54	0	0.5	0	0.1	–	–	–	–	–	–	–
Ukraine	3,102	2,956	3,072	–15	–0.5	23	0.8	1,535	1,799	1,715	26	1.6	–17	–1.0
United Kingdom	1,862	1,914	1,902	5	0.3	–2	–0.1	103	231	286	13	8.4	11	4.4
Total Europe (33 countries)	**54,492**	**60,708**	**64,177**	**622**	**0.0**	**694**	**1.1**	**13,802**	**15,501**	**16,094**	**170**	**0.0**	**119**	**0.8**
North and Central America														
North America														
Country/area														
Canada	3,976	8,147	10,206	417	7.4	412	4.6	–	–	–	–	–	–	–
United States of America	10,305	16,274	17,061	597	4.7	157	0.9	–	–	–	–	–	–	–

(*continued*)

Table A.3. Continued

	Productive							Protective						
	Area ('000 ha)			Annual change rate				Area ('000 ha)			Annual change rate			
	1990	2000	2005	1990–2000 '000 ha/ year	%	2000–2005 '000 ha/ year	%	1990	2000	2005	1990–2000 '000 ha/ year	%	2000–2005 '000 ha/ year	%
Total (2 countries)	**14,281**	**24,421**	**27,267**	**1,014**	**0.0**	**569**	**2.2**	**0**	**0**	**0**	**–**	**–**	**–**	**–**
Total North and Central America (2 countries)	**14,281**	**24,421**	**27,267**	**1,014**	**0.0**	**569**	**2.2**	**0**	**0**	**0**	**–**	**–**	**–**	**–**
Oceania														
Oceania														
Country/area														
Australia	1,023	1,485	1,766	46	3.8	56	3.5	–	–	–	–	–	–	–
New Zealand	1,261	1,767	1,832	51	3.4	13	0.7	–	2	20	0	–	4	58.5
Total Oceania (2 countries)	**2,284**	**3,252**	**3,598**	**97**	**0.0**	**69**	**2.0**	**0**	**2**	**20**	**0**	**–**	**4**	**58.5**
South America														
South America														
Country/area														
Argentina	769	1,078	1,229	31	3.4	30	2.7	–	–	–	–	–	–	–
Brazil	5,070	5,279	5,384	21	0.4	21	0.4	–	–	–	–	–	–	–

Chile	1,766	2,379	2,686	61	3.0	61	2.5	–	–	–	–	–	–	–
Uruguay	197	655	751	46	12.8	19	2.8	4	14	15	1	13.3	0	1.4
Venezuela (Bolivarian Republic of)	873	836	806	–4	–0.4	–6	–0.7	28	27	26	0	–0.4	0	–0.8
Total South America (5 countries)	**8,675**	**10,227**	**10,856**	**155**	**0.0**	**126**	**1.2**	**32**	**41**	**41**	**1**	**0.0**	**0**	**0.0**
WORLD	**151,289**	**178,408**	**197,403**	**2,712**	**0.0**	**3,799**	**2.0**	**49,890**	**56,214**	**62,028**	**632**	**0.0**	**1,163**	**2.0**

Source: FAO Global Planted Forests Thematic Study – Results and Analysis (FAO, 2006c).

Bibliography

ABARE and Jaakko Pöyry (1999) *Global Outlook for Plantations*. ABARE Research Report 99.9, Canberra, Australia.

Aber, J.D. (1987) Restored forests and the identification of critical factors in species-site interactions. In: Jordan III, W.R., Gilpin, M.E. and Aber, J.D. (eds) *Restoration Ecology: A Synthetic Approach to Ecological Research*. Cambridge University Press, Cambridge, pp. 241–250.

Acharya, K.P. (2002) Twenty-four years of community forestry in Nepal. *International Forestry Review* 4, 149–156.

Adlard, P.G. (1993) Historical background. *Study No. 1 Shell/WWF Tree Plantation Review*. Shell International Petroleum Group, London and World Wide Fund for Nature, Godalming.

Alig, R.J. and Butler, B.J. (2004) Projecting large-scale area changes in land use and land cover for terrestrial carbon analyses. *Environmental Management* 33, 443–456.

Alig, R.J., Adams, D.M., Chmelik, J.T. and Bettinger, P. (1999) Private forest investments and long-run sustainable harvest volumes. *New Forests* 17, 301–327.

Alvarado, L.X.R. and Wertz-Kanounnikoff, S. (2007) Why are we seeing 'REDD'. An analysis of the international debate on reducing emissions from deforestation and degradation in developing countries. *Analyses No. 2/2007 Natural Resources*. Institut du Développement durable et des Relations Internationales, Paris.

Ambar, S. (1986) Conversion of forest lands to annual crops: an Indonesian perspective. In: *Land Use, Watersheds, and Planning in the Asia-Pacific Region*. FAO RAPA report 1986/3. FAO, Bangkok, pp. 95–111.

Anon. (1923) Observations on the results of artificial forestry in Germany. *Journal of Forestry* 21, 718–722.

Anon. (1999) The Role of Planted Forests in Sustainable Forest Management. Intersessional Experts Meeting, UN Inter-governmental Forum on Forests, Santiago, Chile.

Anon. (2003) The Role of Planted Forests in Sustainable Forest Management. UNFF Intersessional Experts Meeting, Wellington, New Zealand.

Armesto, J.J., Smith-Ramirez, C. and Rozzi, R. (2001) Conservation strategies for biodiversity and indigenous people in Chilean forest ecosystems. *Journal of the Royal Society of New Zealand* 31, 865–877.

Ashby, W.C. and Vogel, W.G. (1993) *Tree Planting on Minelands in the Midwest: A Handbook*. Coal Research Centre, Southern University of Illinois, Carbondale, Illinois.

Attiwill, P.M. and Weston, C.J. (2001) Forest soils. In: Evans, J. (ed.) *The Forests Handbook. Vol. 1 An Overview of Forest Science*. Blackwell Science, Oxford, pp. 157–187.

Bailey, C., Sinclair, P.R. and Dubois, M.R. (2004) Future forests: forecasting social and ecological consequences of genetic engineering. *Society & Natural Resources* 17, 641–650.

Barbour, J. (2004) Wood quality. In: Burley, J., Evans, J. and Youngquist, J. (eds) *Encyclopedia of Forest Science*. Elsevier, Oxford, pp. 1840–1845.

Barlow, K. (2003) Reconstructing rurality and community: plantation forestry in Victoria, Australia. *Journal of Rural Studies* 19, 503–519.

Bass, S. (2004) Certification. In: Burley, J., Evans, J. and Youngquist, J. (eds) *Encyclopedia of Forest Science*. Elsevier, Oxford, pp. 1350–1356.

Baumer, M. (1990) *Agroforestry and Desertification*. Technical Centre of Agricultural and Rural Cooperation, Wageningen.

Bednall, B.H. (1968) The problem of lower volumes associated with second rotations in *Pinus radiata* plantations in South Australia. Paper presented at the 9th Commonwealth Forestry Conference, India.

Belcher, B., Michon, G., Angelsen, A., Perez, M.R. and Asbjornsen, H. (2005) The socio-economic conditions determining the development, persistence, and decline of forest garden systems. *Economic Botany* 59, 245–253.

Ben Salem, B. (1991) Prevention and control of wind erosion in arid regions. *Unasylva* 42 (164), 33–39.

Bene, J.G., Beall, H.W. and Coté, A. (1977) *Trees, Food, and People: Land Management in the Tropics*. International Development Research Centre, Ottawa.

Bernard-Reversat, F. (ed.) (2001) *Effect of Exotic Tree Plantations on Plant Diversity and Biological Soil Fertility in Congo Savanna: With Special Reference to Eucalypts*. Center for International Forestry Research, Bogor, Indonesia.

Berndes, G., Hoogwijk, M. and van den Broek, R. (2003) The contribution of biomass in the future global energy supply: a review of 17 studies. *Biomass & Bioenergy* 25, 1–28.

Berthe, Y. (1997) The role of forestry in combating desertification. In: *Protective and Environmental Functions of Forests. Proceedings of the XI World Forestry Congress*, 13–22 October 1997, Antalya, Turkey. Ministry of Forestry, Ankara, Vol. 2, pp. 307–311.

Bertomeu, M., Bertomeu, M. and Gimenez, J.C. (2006) Improving adoptability of farm forestry in the Philippine uplands: a linear programming model. *Agroforestry Systems* 68, 81–91.

Betts, M.G., Diamond, A.W., Forbes, G.J., Frego, K., Loo, J.A., Matson, B. and Roberts M.R. (2005) Plantations and biodiversity: a comment on the debate in New Brunswick. *Forestry Chronicle* 81, 265–269.

Bhat, D.M., Murali, K.S. and Ravindranath, N.H. (2001) Formation and recovery of secondary forests in India: a particular reference to Western Ghats in South India. *Journal of Tropical Forest Science* 13, 601–620.

Binkley, C.S. (1999) Ecosystem management and plantation forestry: new directions in British Columbia. *New Forests* 18, 75–88.

Binkley, C.S., Aronow, M.E., Washburn, C.L. and New, D. (2005) Global perspectives on intensively managed plantations: implications for the Pacific Northwest. *Journal of Forestry* 103, 61–64.

Binns, W.O. (1962) Some aspects of peat as a substrate for tree growth. *Irish Forestry* 19, 32–55.

Bloomfield, J. (2000) Land use, land-use change, forestry, and agricultural activities in the clean development mechanism: estimates of greenhouse gas offset potential. *Mitigation and Adaptation Strategies for Global Change* 5, 9.

Boardman, R. (1988) Living on the edge – the development of silviculture in South Australian pine plantations. *Australian Forestry* 51, 135–156.

Bollen, A. and Donati, G. (2006) Conservation status of the littoral forest of south-eastern Madagascar: a review. *Oryx* 40, 57–66.

Bonneau, M., Gelpe, J. and Le Tacon, F. (1968) Influence of mineral nutrition on the dying of *Pinus pinaster* in the Landes in Gascony. *Annales Science Forestiere, Paris* 25, 251–289.

Boston, K. and Sessions, J. (2006) Development of a spatial harvest scheduling system to promote the conservation between indigenous and exotic forests. *International Forestry Review* 8, 297–306.

Bowyer, J.L. (2004) Environmental benefits of wood as a building material. In: Burley, J., Evans, J. and Youngquist, J. (eds) *Encyclopedia of Forest Science*. Elsevier, Oxford, pp. 1863–1869.

Boyle, J.R., Winjum, J.K., Kavanagh, K. and Jensen, E.C. (eds) (1999) *Planted Forests: Contributions to the Quest for Sustainable Societies*. Kluwer Academic, Dordrecht.

Bradshaw, A.D. (1987) Restoration: an acid test for ecology. In: Jordan III, W.R., Gilpin, M.E. and Aber, J.D. (eds) *Restoration Ecology: A Synthetic Approach to Ecological Research*. Cambridge University Press, Cambridge, pp. 23–29.

Brazier, C.M., Jung, T. and Oswald, W. (2006) Progress in Research on Phytophthora Diseases of Forest Trees. *Proceedings of the Third International IUFRO Working Party S07.02.09*, Freising, Germany 2004. Forests Research, Farnham.

Briggs, R.D. (2004) The forest floor. In: Burley, J., Evans, J. and Youngquist, J. (eds) *Encyclopedia of Forest Science*, Vol. 3. Elsevier, Oxford, pp. 1223–1227.

Broadmeadow, M. and Ray, D. (2005) Climate change and British woodland. *Forestry Commission Information Note FCIN069*. Forestry Commission, Edinburgh.

Broun, A.F. (1912) *Silviculture in the Tropics*. Macmillan, London.

Brown, S. and Lugo, A.E. (1994) Rehabilitation of tropical lands: a key to sustaining development. *Restoration Ecology* 2, 97–101.

Bruijnzeel, L.A. (1997) Hydrology of forest plantations in the tropics. In: Nambiar, E.K.S. and Brown, A.G. (eds) *Management of Soil, Nutrients and Water in Tropical Plantation Forests*. ACIAR Monograph No. 43. Australian Centre for International Agriculture Research, CSIRO Australia and Center for International Forestry Research, Indonesia, pp. 125–167.

Bruijnzeel, L.A. (2004) Hydrological cycle. In: Burley, J., Evans, J. and Youngquist, J. (eds) *Encyclopedia of Forest Science*. Elsevier, Oxford, pp. 340–350.

Brundtland, G.H. (1987) *Our Common Future*. Report of the World Commission on Environment and Development, Oxford.

Burdon, R.D. (2004) Forest genetics and tree breeding: current and future signposts. In: Burley, J., Evans, J. and Youngquist, J. (eds) *Encyclopedia of Forest Science*, Vol. 4. Elsevier, Oxford, pp. 1538–1545.

Burger, J.A. (2004) Soil and its relationship to forest productivity and health. In: Burley, J., Evans, J. and Youngquist, J. (eds) *Encyclopedia of Forest Science*, Vol. 3. Elsevier, Oxford, pp. 1189–1194.

Butterfield, R.P. (1995) Promoting biodiversity: advances in evaluating native species for reforestation. *Forest Ecology and Management* 75, 111–121.

Byrne, K.A. and Farrell, E.P. (2005) The effect of afforestation on soil carbon emissions in blanket peats in Ireland. *Forestry* 78, 209–216.

Byrne, K.A. and Milne, R. (2006) Carbon stocks and sequestration in plantation forests in the Republic of Ireland. *Forestry* 79, 361–369.

Calder, I.R. (1992) Water use of eucalypts – a review. In: Calder, I.R., Hall, R.L. and Adlard, P.G. (eds) *Growth and Water Use of Forest Plantations*. Wiley and Sons, New York, pp. 167–179.

Calder, I.R. (1996) Water use by forests at the plot and catchment scale. *Commonwealth Forestry Review* 75, 19–30.

Calder, I.R. (2002) Eucalyptus, water and the environment. In: Coppen, J.J.W. (ed.) *Eucalyptus: The Genus Eucalyptus*. Taylor and Francis, London, pp. 36–51

Calder, I.R., Reid, I., Nisbet, T.R., Robinson, M. and Walker, D. (1997) *TADPOLE: Trees and drought project in lowland England. Scoping study*. Report to Department of the Environment, London, March 1997.

Calder, I.R., Reid, I., Nisbet, T.R. and Green, J.C. (2003) Impact of lowland forests in England on water resources: application of the Hydrological Land Use Change (HYLUC) model. *Water Resources Research* 39, 1319.

Cameron, A.D., Mason, W.L. and Malcolm, D.C. (eds) (2001) Transformation of plantation forests. Selected and edited papers of IUFRO conference, Edinburgh, August 1999. *Forest Ecology and Management* 151 special issue.

Campinhos, E. (1999) Sustainable plantations of high yield *Eucalyptus* trees for production of fibre: the Aracruz case. *New Forests* 17, 129–143.

Cannell, M.G.R. (2002) Impacts of climate change on forest growth. In: Broadmeadow, M. (ed.) *Climate Change: Impacts on UK Forests. Forestry Commission Bulletin 125*. Forestry Commission, Edinburgh, pp. 141–148.

Cannell, M.G.R. (2004) Short rotation forestry for biomass production. In: Burley, J., Evans, J. and Youngquist, J. (eds) *Encyclopedia of Forest Science*. Elsevier, Oxford, pp. 872–877.

Cannell, M.G.R., Dewar, R.C. and Pyatt, D.G. (1993) Conifer plantations on drained peatland in Britain: a net gain or loss of carbon. *Forestry* 66, 353–369.

Cannell, M.G.R., Thornley, J.H.M., Mobbs, D.C. and Friend, A.D. (1998) UK conifer forests may be growing faster in response to N deposition, atmospheric CO_2 and temperature. *Forestry* 71, 277–296.

Carle, J.C. (2007) Vulnerabilities of smallholder plantings. *Unasylva* 58 (228), 3.

Carle, J.C. and Holmgren, P. (2003) Definitions related to planted forests. Paper 22, UNFF Intersessional Experts Meeting on the Role of Planted Forests in Sustainable Management, 24–30 March 2003, Wellington, New Zealand.

Carle, J.C. and Holmgren, P. (2008) Wood from Planted Forests: A Global Outlook 2005-2030. *Forest Products Journal* 58 (12), 6-18.

Carle, J.C. and Ma, Q. (2005) Challenges of translating science into practice: poplars and other species in the Three North Region of China. *Unasylva* 56 (2), 31-37.

Carnevale, N.J. and Montagnini, F. (2002) Facilitating regeneration of secondary forests with the use of mixed and pure plantations of indigenous tree species. *Forest Ecology and Management* 163, 217–227.

Carnus, J.-M., Parrotta, J., Brockerhoff, E., Arbez, M., Jactel, H., Kremer, A., Lamb, D., O'Hara, K. and Walters, B. (2006) Planted forests and biodiversity. *Journal of Forestry* 104, 65–77.

Cavendish, W. (2001) Non-timber forest products and rural poverty: an economic analysis. In: Evans, J. (ed.) *The Forests Handbook. Vol. 1 An Overview of Forest Science*. Blackwell Science, Oxford, pp. 372–391.

Cerli, C., Celi, L., Johansson, M.B., Kogel-Knabner, I., Rosenqvist, L. and Zanini, E. (2006) Soil organic matter changes in a spruce chronosequence on Swedish former agricultural soil I. Carbon and lignin dynamics. *Soil Science* 171, 837–849.

Chacko, K.C. (1995) Silvicultural problems in management of teak plantations. *Proceedings of the 2nd Regional Seminar on Teak*, 'Teak for the Future', Yangon, Myanmar, May 1995. FAO (Bangkok), pp. 91–98.

Chen, G.S., Yang, Y.S., Xie, J.S., Guo, J.F., Gao, R. and Qian, W. (2005) Conversion of a natural broad-leafed evergreen forest into pure plantation forests in a subtropical area: effects on carbon storage. *Annals of Forest Science* 62, 659–668.

Chundamannii, M. (1998) *Teak Plantations in Nilambur – An Economic Review*. KFRI Research Report No. 144. Kerala Forest Research Institute, Peechi.

Ciesla, W.M. and Donaubauer, E. (1994) Decline and Dieback of Trees and Forests. *FAO Forestry Paper 120*. FAO, Rome.

Clark, J. (2001) The global wood market, prices and plantation investment: an examination drawing on the Australian experience. *Environmental Conservation* 28, 53–64.

Clay, D.V., Dixon, F.L. and Willoughby, I. (2005) Natural products as herbicides for tree establishment. *Forestry* 78, 1–10.

Cook, J. (2000) Bioenergy in the United States: progress and possibilities. *Biomass & Bioenergy* 18, 441–455.

Cossalter, C., Pye-Smith, C. and Kaimowitz, D. (2003) *Fast Wood Forestry – Good or Bad Land Use*. Center for International Forestry Research, Bogor.

Craib, I.J. (1934) The place of thinning in wattle silviculture and its bearing on the management of exotic conifers. *Zeitschrift für die Welt forstwirtschaft* 1, 77–108.

Craib, I.J. (1947) Silviculture of exotic conifers in South Africa. Paper presented at the 5th British Empire Forestry Conference, South Africa.

Dawson, J.J.C. and Smith, P. (2007) Carbon losses from soil and its consequence for land-use management. *Science of the Total Environment* 382, 165–190.

de Jong, W., Chokkalingam, U., Smith, J. and Sabogal, C. (2001) Tropical secondary forests in Asia: introduction and synthesis. *Journal of Tropical Forest Science* 13, 563–576.

Denyer, K., Burns, B. and Ogden, J. (2006) Buffering of native forest edge microclimate by adjoining tree plantations. *Austral Ecology* 31, 478–489.

Dickmann, D.I. (2006) Silviculture and biology of short-rotation woody crops in temperate regions: then and now. *Biomass & Bioenergy* 30, 696–705.

Dimitriou, I. and Aronsson, P. (2005) Willow for energy and phytoremediation. *Unasylva* 56 (2), 47-50.

Ding, Y.X. and Chen, J.L. (1995) Effect of continuous plantation of Chinese fir on soil fertility. *Pedosphere* 5, 57–66.

Dlomo, M. and Pitcher, M. (2005) Juggling social and economic goals: South Africa. In: Garforth, M. and Mayers, J. (eds) *Plantations, Privatisation, Poverty and Power – Changing Ownership and Management of State Forests*. Earthscan, London, pp. 223–245.

Durst, P.B., Waggener, T.R., Enters, T. and Tan, L.C. (eds) (2001) *Forests Out of Bounds: Impacts and Effectiveness of Logging Bans in Natural Forests in Asia-Pacific*. RAP Publication 2001/08. Food and Agriculture Organization of the United Nations, Regional Office for Asia and the Pacific, Bangkok.

Durst, P.B., McKenzie, P.J., Brown, C.L. and Appanah, S. (2006) Challenges facing certification and eco-labelling of forest products in developing countries. *International Forestry Review* 8, 193–200.

Dutschke, M. and Wolf, R. (2007) *Reducing Emissions from Deforestation in Developing Countries – The Way Forward*. GTZ, Eschborn.

Dyck, W.J. and Skinner, M.F. (1990) Potential for productivity decline in New Zealand radiata pine forests. In: Gessel, S.P., Lacate, D.S., Weetman, G.F. and Powers, R.F. (eds) *Sustained Productivity of Forest Soils. Proceedings of the 7th North American Forest*

Soils Conference. University of British Columbia, Faculty of Forestry Publication, Vancouver, pp. 318–332.

Dyck, W.J., Cole, D.W. and Comerford, N.B. (1994) *Impacts of Forest Harvesting on Long-term Site Productivity*. Chapman and Hall, London.

Dye, P.J. (1996) Climate, forest and streamflow relationships in South African afforested catchments. *Commonwealth Forestry Review* 75, 31–36.

Earl, D.E. (1975) *Forest Energy and Economic Development*. Clarendon Press, Oxford.

Elauria, J.C., Castro, M.L.Y. and Racelis, D.A. (2003) Sustainable biomass production for energy in the Philippines. *Biomass & Bioenergy* 25, 531–540.

Elfling, B. and Nystrom, K. (1996) Yield capacity of planted *Picea abies* in northern Sweden. *Scandinavian Journal of Forestry Research* 11, 38–49.

Er, K.B.H. and Innes, J.L. (2003) The presence of old-growth characteristics as a criterion for identifying temperate forests of high conservation value. *International Forestry Review* 5, 1–8.

Espinosa, M., Acuna, E., Cancino, J., Munoz, F. and Perry, D.A. (2005) Carbon sink potential of radiata pine plantations in Chile. *Forestry* 78, 11–20.

Evans, H.F. (2001) Management of pest threats. In: Evans, J. (ed.) *The Forests Handbook. Vol. 2 Applying Forest Science for Sustainable Management*. Blackwell Science, Oxford, pp. 172–201.

Evans, H.F., Straw, N. and Watt, A. (2002) Climate change: impacts for insect pests. In: Broadmeadow, M. (ed.) *Climate Change: Impacts on UK Forests. Forestry Commission Bulletin 125*. Forestry Commission, Edinburgh, pp. 99–118.

Evans, J. (1976) Plantations: productivity and prospects. *Australian Forestry* 39, 150–163.

Evans, J. (1978) A further report on second rotation productivity in the Usutu Forest, Swaziland – results of the 1977 reassessment. *Commonwealth Forestry Review* 57, 253–261.

Evans, J. (1984) Measurement and prediction of changes in site productivity. In: Grey, D.C., Schonau, A.P.G., Schutz, C.J. and van Laar, A. (eds) *IUFRO Symposium on Site and Productivity of Fast Growing Plantations*. Pretoria and Pietermaritzburgh, South Africa April 1984, pp. 970–920.

Evans, J. (1999) *Sustainability of forest plantations – the evidence*. Issues paper. UK Department for International Development, London.

Evans, J. (2001a) How to be successful in plantation development. *Tropical Forest Update* 11, 3–5. The International Tropical Timber Organisation, Yokohama, Japan.

Evans, J. (2001b) *Biological Sustainability of Productivity in Successive Rotations*. Forest Plantations Thematic Working Paper FP/2, edited by D.J. Mead. Forest Department, FAO, Rome.

Evans, J. (2004a) Sustainability of forest plantations. In: Burley, J., Evans, J. and Youngquist, J. (eds) *Encyclopedia of Forest Science*, Vol. 2. Elsevier, Oxford, pp. 865–872.

Evans, J. (2004b) Forest plantations. In: Burley, J., Evans, J. and Youngquist, J. (eds) *Encyclopedia of Forest Science*, Vol. 2. Elsevier, Oxford, pp. 822–828.

Evans, J. (2005) Growth rates over four rotations of pine in Swaziland. *International Forestry Review* 7, 305–310.

Evans, J. and Hibberd, B.G. (1990) Managing to diversify forests. *Aboricultural Journal* 14, 373–378.

Evans, J. and Masson, P. (2001) Sustainable plantation forestry: a case study of wood production and environmental management strategies in the Usutu Forest, Swaziland. In: Evans, J. (ed.) *The Forests Handbook. Vol. 2 Applying Forest Science for Sustainable Management*. Blackwell Science, Oxford, pp. 357–374.

Evans, J. and Turnbull, J. (2004) *Plantation Forestry in the Tropics*, 3rd edn. Oxford University Press, Oxford.

Evans, J. and Wright, D. (1988) The Usutu Pulp Company – development of an integrated forestry project. In: Carr M. (ed.) *Sustainable Industrial Development. Seven Case Studies*. Intermediate Technology Publications, London, pp. 151–171.

Ewers, R.M., Kliskey, A.D., Walker, S., Rutledge, D., Harding, J.S. and Didham, R.K. (2006) Past and future trajectories of forest loss in New Zealand. *Biological Conservation* 133, 312–325.

Fahey, B. and Jackson, R. (1995) Hydrological impacts of converting native forests and grasslands to pine plantation, South Island, New Zealand. In: Korpilahti, E., Salonen, T. and Oja, S. (eds) *Caring for the Forest: Research in a Changing World*. Congress Report Abstracts of Invited Papers, IUFRO XX World Congress, 6–12 August 1995, Tampere, Finland. The Finnish IUFRO World Congress Organising Committee, Tampere, Finland, p. 31.

FAO (1967) Actual and potential role of man-made forests in the changing world pattern of wood consumption. Secretariat Note. *Proceedings of the World Symposium on Man-made Forests and their Industrial Importance*. Food and Agriculture Organization of the United Nations, Rome, pp. 1–50.

FAO (1978) *Forestry for Rural Communities*. Food and Agriculture Organization of the United Nations, Rome.

FAO (1988) *An Interim Report on the State of Forest Resources in the Developing Countries*. FO; MISC/88/7 Food and Agriculture Organization of the United Nations, Rome.

FAO (1992) Mixed and pure forest plantations in the tropics and subtropics. *FAO Forestry Paper 103*. FAO, Rome.

FAO (1993) *Forest Resources Assessment 1990*. FAO, Rome.

FAO (2000) The global outlook for future wood supply from forest plantations. *Working Paper GFPOS/WP/03*. FAO, Rome.

FAO (2001) Global forest resources assessment 2000. Main report. *FAO Forestry Paper 140*. Food and Agriculture Organization of the United Nations, Rome.

FAO (2003) Harmonizing forest-related definitions for use of various stakeholders. Second Expert Meeting, 11–13 September 2002. Food and Agriculture Organization of the United Nations, Rome.

FAO (2005a) *Global Forest Resources Assessment 2005*. UN Food and Agriculture Organization, Rome.

FAO (2005b) *State of the World's Forests 2005*. UN Food and Agriculture Organization, Rome.

FAO (2006a) Global forest resources assessment 2005 – progress towards sustainable forest management. *FAO Forestry Paper 147*. UN Food and Agriculture Organization, Rome.

FAO (2006b) *Responsible Management of Planted Forests: Voluntary Guidelines* Planted Forests and Trees Working Papers FP/37E FAO, Rome.

FAO (2006c) *Global planted forests thematic study: results and analysis* by A. Del Lungo, J. Ball and J. Carle. Planted Forests and Trees Working Paper 38, FAO, Rome.

FAO (2008a) *Forests and Energy – key issues*. FAO Forestry Paper 154, Rome.

FAO (2008b) *Forests and Water* by L.S. Hamilton. FAO Forestry Paper 155, Rome.

FAO (2009) *Global Review of Forest Pests and Diseases*. FAO Forestry Paper 156, Rome.

Farley, K.A., Jobbagy, E.G. and Jackson, R.B. (2005) Effects of afforestation on water yield: a global synthesis with implications for policy. *Global Change Biology* 11, 1565–1576.

Fernandez, E.B. (1997) Strategies for strengthening watershed management in tropical mountain areas. In: *Protective and Environmental Functions of Forests. Proceedings of the XI World Forestry Congress*, 13–22 October 1997, Antalya, Turkey. Ministry of Forestry, Ankara, Vol. 2, pp. 247–255.

Ferris-Kaan, R. (1995) Managing forests for biodiversity. *Forestry Commission Technical Paper 8*. Forestry Commission, Edinburgh.

Ferris, R. and Carter, C. (2000) Managing rides, roadsides and edge habitats in lowland forests. *Forestry Commission Bulletin 123*. Forestry Commission, Edinburgh.

Fimbel, R.A. and Fimbel, C.C. (1996) The role of exotic conifer plantations in rehabilitating degraded forest lands: a case study from Kibale Forest in Uganda. *Forest Ecology and Management* 81, 215–226.

Finlayson, W. (1998) *Effects of Deforestation and of Tree Planting on the Hydrology of the Upper Mahaweli Catchment: A Review of Published Evidence*. Environment and Forest Conservation Division, Mahaweli Authority of Sri Lanka.

Florence, R.G. (1996) *Ecology and Silviculture of Eucalypt Forests*. CSIRO, Australia.

Folster, H. and Khanna, P.K. (1997) Dynamics of nutrient supply in plantation soils. In: Nambiar, Sadanandan, E.K. and Brown, A.G. (eds) *Management of Soil, Nutrients and Water in Tropical Plantation Forests*. ACIAR (Australian Centre for International Agricultural Research) Monograph No. 43. 339–378.

Forest Restoration Research Unit (2000) *Tree Seeds and Seedlings for Restoring Forests in Northern Thailand*. Biology Department, Chiang Mai University, Chiang Mai.

Forestry Commission (2007) *Forestry Facts and Figures 2007*. Forestry Commission, Edinburgh.

Fox, D. (2007) Saved by the trees? *New Scientist* 196 (27 October 2007), 42–46.

Fox, T.R. (2000) Sustained productivity in intensively managed forest plantations. *Forest Ecology and Management* 138, 187–202.

Fox, T.R., Jokela, E.J. and Allen, H. Lee (2007) The development of pine plantation silviculture in the southern United States. *Journal of Forestry* 105, 337–347.

Fung, P.Y.H., Kirschbaum, M.U.F., Raison, R.J. and Stucley, C. (2002) The potential for bioenergy production from Australian forests, its contribution to national greenhouse targets and recent developments in conversion processes. *Biomass & Bioenergy* 22, 223–236.

Garcia-Quijano, J.F., Deckmyn, G., Moons, E., Proost, S., Ceulemans, R. and Muys, A. (2005) An integrated decision support framework for the prediction and evaluation of efficiency, environmental impact and total social cost of domestic and international forestry projects for greenhouse gas mitigation: description and case studies. *Forest Ecology and Management* 207, 245–262.

Gardner, B. (2004) Windbreaks and shelterbelts. In: Burley, J., Evans, J. and Youngquist, J. (eds) *Encyclopedia of Forest Science*, Vol. 4. Elsevier, Oxford, pp. 1801–1805.

Gardner, J. (2001) Rehabilitating mines to meet land use objectives after bauxite mining I. The jarrah forests of Western Australia. *Unasylva* 207 (52), 3–8.

Garforth, M. and Mayers, J. (eds) (2005) *Plantations, Privatisation, Poverty and Power – Changing Ownership and Management of State Forests*. Earthscan, London.

Garforth, M., Landell-Mills, N. and Mayers, J. (2005a) Plantations, livelihoods and poverty. In: Garforth, M. and Mayers, J. (eds) *Plantations, Privatisation, Poverty and Power – Changing Ownership and Management of State Forests*. Earthscan, London, pp. 3–21.

Garforth, M., Landell-Mills, N. and Mayers, J. (2005b) Conclusions and ways forward. In: Garforth, M. and Mayers, J. (eds) *Plantations, Privatisation, Poverty and Power – Changing Ownership and Management of State Forests*. Earthscan, London, pp. 90–98.

Garten, C.T. (2002) Soil carbon storage beneath recently established tree plantations in Tennessee and South Carolina, USA. *Biomass & Bioenergy* 23, 93–102.

Geldenhuys, C.J. (1997) Native forest regeneration in pine and eucalypt plantations in Northern Province, South Africa. *Forest Ecology and Management* 99, 101–115.

Georgaidis, N.M. and Cooper, R.J. (2007) Development of a forest certification standard compatible with PEFC and FSC's management requirements. A case study from Greece. *Forestry* 80, 113–135.

Ghazoul, J. and Evans, J. (2004) Causes of deforestation and forest fragmentation. In: Burley, J., Evans, J., and Youngquist, J. (eds) *Encyclopedia of Forest Science*. Elsevier, Oxford, Vol. 3, pp. 1367–1375.

Ghosh, R.C. (1978) Evaluating and analysing environmental impacts of forests in India. *Proceedings of the 8th World Forestry Congress*, Jakarta. FAO, Rome, Vol. 7A, pp. 475–484.

Gibson, A. and Bachelard, E.P. (1994) Relationships between site characteristics and survival strategies of *Eucalyptus camaldulensis* seedlings. In: Brown, A.G. (ed.) *Australian Tree Species Research in China. ACIAR Proceedings No. 48.* Australian Centre for International Agricultural Research, Canberra, pp. 91–95.

Gibson, A., Bachelard, E.P. and Hubick, K.T. (1995) Relationship between climate and provenance variation in *Eucalyptus camaldulensis* Dehnh. *Australian Journal of Plant Physiology* 22, 453–460.

Gibson, I.A.S. and Jones, T. (1977) Monoculture as the origin of major forest pests and diseases. In: Cherrett, J.M. and Sagar, G.R. (eds) *Origins of Pest, Parasite, Disease and Weed Problems*. Blackwell, Oxford, pp. 139–161.

Gill, G. and McIntosh, R. (2001) Restructuring of plantation forest: Kielder, United Kingdom. In: Evans, J. (ed.) *The Forests Handbook. Vol. 2 Applying Forest Science for Sustainable Management*. Blackwell Science, Oxford, pp. 298–309.

Gonçalves, J.L.M., Barros, N.F., Nambiar, E.K.S. and Novais, R.F. (1997) Soil and stand management for short-rotation plantations. In: Nambiar, Sadanandan, E.K. and Brown, A.G. (eds) *Management of soil, nutrients and water in tropical plantation forests*. ACIAR (Australian Centre for International Agricultural Research) Monograph No. 43, 379–417.

Gonçalves, J.L. de M., Benedetti, V. and McNabb, K. (eds and translator) (2004a) *Forest Nutrition and Fertilization*. Institute of Forest Research and Study (IPEF) Piracicaba, Sao Paulo.

Gonçalves, J.L. de M., Stape, J.L., Benedetti, V., Fessel, V.A.G. and Gava, J.L. (2004b) An evaluation of minimum and intensive soil preparation regarding fertility and tree nutrition. In: Gonçalves, J.L. de M., Benedetti, V. and McNabb, K. (eds and translator) *Forest Nutrition and Fertilization*. Institute of Forest Research and Study (IPEF) Piracicaba, Sao Paulo, pp. 13–64.

Goodall, J. and Klein, H. (2000) Invader plant control in forestry. In: Owen, D.L. (ed.) *South African Forestry Handbook*. South African Institute of Forestry, Pretoria, pp. 253–264.

Göttle, A. and Sene, E.M. (1997) Protective and environmental functions of forests. In: *Protective and Environmental Functions of Forests. Proceedings of the XI World Forestry Congress*, 13–22 October 1997, Antalya, Turkey. Ministry of Forestry, Ankara, Vol. 2, pp. 233–243.

Gouldin, C.J. (2006) Forest stewardship council certification of industrial plantation forests. *Allgemeine Forst Und Jagdzeitung* 177, 42–47.

Greig, S. (2007) Carbon issues in UK forestry. *Quarterly Journal of Forestry* 101, 128–134.

Griffith, A.L. (1942) Teak plantation technique. *Indian Forest Records* 5, 123–219.

Griffith, A.L. and Gupta, R.S. (1948) Soils in relation to teak with special reference to laterisation. *Indian Forestry Bulletin No. 141*.

Grundy, D.S. (2005) From plantation developer to steward of the nation's forests: the UK. In: Garforth, M. and Mayers, J. (eds) *Plantations, Privatization, Poverty and Power: Changing Ownership and Management of State Forests.* Earthscan, London, pp. 246–265.

Guo, L.B. and Gifford, R.M. (2002) Soil carbon stocks and land use change: a meta analysis. *Global Change Biology* 8, 345–360.

Hall, J.P. and Richardson, J. (2001) ENFOR – Energy from the forest. *Forestry Chronicle* 77, 831–835.

Hardwick, K., Healey, J., Elliott, S., Garwood, N. and Anusarnsunthorn, V. (1997) Understanding and assisting natural regeneration processes in degraded seasonal evergreen forests in northern Thailand. *Forest Ecology and Management* 99, 203–214.

Hargreaves, K.A., Milne, R. and Cannell, M.G.R. (2003) Carbon balance of afforested peatland in Scotland. *Forestry* 76, 299–316.

Harmer, R. and Kiewett, A. (2006) Restoration of lowland conifers PAWS. *Forest Research Annual Report 2005–2006*. Forestry Commission, Edinburgh, pp. 24–29.

Harmer, R., Thompson, R. and Humphrey, J. (2005) Great Britain – conifers to broadleaves. In: Stanturf, J.A. and Marsden, P. (eds) *Restoration of Boreal and Temperate Forests*. CRC Press, Boca Raton, pp. 319–338.

Hayes, J.P., Schoenholtz, S.H., Hartley, M.J., Murphy, G., Powers, R.F., Berg, D. and Radosevich, R. (2005) Environmental Consequences of Intensively Managed Forest Plantations in the Pacific Northwest. *Journal of Forestry* 103, 83–87.

Heriyanto, N.M., Heriansyah, I., Siregar, C.A. and Kiyoshi, M. (2003) Potential of *Acacia mangium* plantations for carbon sequestration. Paper presented at Advances in Genetic Improvement of Tropical Tree Species. *Proceedings of the International Conference*, Yogyakarta, Indonesia, 1–3 October, 2002.

Hewlett, J.D. (1982) *Principles of Forest Hydrology*. University of Georgia Press, Athens.

Hibberd, B.G. (1985) Restructuring of plantations in Kielder Forest District. *Forestry* 58, 119–129.

Higman, S., Mayers, J., Bass, S., Judd, N. and Nussbaum, R. (2005) *The Sustainable Forestry Handbook: A Practical Guide for Tropical Forest Managers on Implementing New Standards*, 2nd edn. Earthscan, London.

Hoffmann, D. and Weih, M. (2005) Limitations and improvement of the potential utilisation of woody biomass for energy derived from short rotation woody crops in Sweden and Germany. *Biomass & Bioenergy* 28, 267–279.

Holmsgaard, E., Holstener-Jorgensen, H. and Yde-Andersen, A. (1961) Bodenbildung, Zuwachs und Gesundheitszustand von Fichtenbestanden erster und sweiter Generation. 1. Nord-Zeeland. *Det forstlige Forsogsvaesen i Danmark* 27, 1–67.

Hooda, N. and Rawat, V. (2006) Role of Bio-energy plantations for carbon-dioxide mitigation with special reference to India. *Mitigation and Adaptation Strategies for Global Change* 11, 437–459.

Houghton, R.A. (2003) Revised estimates of the annual net flux of carbon to the atmosphere from changes in land use and land management 1850–2000. *Tellus Series B – Chemical and Physical Meteorology* 55, 378–390.

Houghton, R.A. and Hackler, J.L. (2003) Sources and sinks of carbon from land-use change in China. *Global Biogeochemical Cycles* 17, no. 2.

Houghton, R.A. and Hackler, J.L. (2006) Emissions of carbon from land use change in sub-Saharan Africa. *Journal of Geophysical Research – Biogeosciences* 111, no. G2.

Howe, G.T., Shindler, B., Cashore, B., Hansen, E., Lach, D. and Armstrong, W. (2005) Public influences on plantation forestry. *Journal of Forestry* 103, 90–94.

Hughes, C.E. (1994) Risks of species introduction in tropical forestry. *Commonwealth Forestry Review* 73, 243–252.

Hulme, M. (1996) *Climate Change and Southern Africa: An Exploration of Some Potential Impacts and Implications in the SADC Region*. Climate Research Unit, University of East Anglia, UK and WWF International.

Humphrey, J.W. (2005) Benefits to biodiversity from developing old-growth conditions in British upland spruce plantations. *Forestry* 78, 33–53.

Huttel, C. and Loumeto, J.J. (2001) Effect of exotic plantations and site management on plant diversity. In: Bernard-Reversat, F. (ed.) *Effect of Exotic Tree Plantations on Plant Diversity and Biological Soil Fertility in Congo Savanna: With Special Reference to Eucalypts*. Center for International Forestry Research, Bogor, pp. 9–18.

Ignaciuk, A.M. and Dellink, R.B. (2006) Biomass and multi-product crops for agricultural and energy production – an AGE analysis. *Energy Economics* 28, 308–325.

IIED (1996) *Towards a Sustainable Paper Cycle*. International Institute for Environment and Development, London.

Innes, J.L. (2004) Carbon cycle. In: Burley, J., Evans, J. and Youngquist, J. (eds) *Encyclopedia of Forest Science*, Vol. 1. Elsevier, Oxford, pp. 139–144.

Innes, J.L. and Hickey, G.M. (2006) The importance of climate change when considering the role of forests in the alleviation of poverty. *International Forestry Review* 8, 406–416.

Ioras, F. and Abrudan, I.V. (2006) The Romanian forestry sector: privatisation facts. *International Forestry Review* 8, 361–365.

ITTO (2007) *ITTO Bi-monthly report*. International Tropical Timber Organisation, Japan.

Jacobs, M.R. (1981) *Eucalypts for Planting*. FAO Forestry Series No. 11. Food and Agriculture Organization of the United Nations, Rome.

Johnson, B. and Kirby, K. (2004) Potential impacts of genetically modified trees on biodiversity of forestry plantations: a global perspective. In: Strauss, S.H. and Bradshaw, H.D. (eds) *The Bioengineered Forest: Challenges for Science and Society*. Resources for the Future, Washington, pp. 190–207.

Johnson, D.W. and Todd, D.E. (1990) Nutrient cycles in forests of Walker Beach Watershed, Tennessee: roles of uptake and leaching causing soil change. *Journal of Environmental Quality* 19, 97–104.

Jones, E.W. (1965) Pure conifers in central Europe. *Journal of Oxford University Forestry Society* 13, 3–15.

Jorgensen, U., Dalgaard, T. and Kristensen, E.S. (2005) Biomass energy in organic farming – the potential role of short rotation coppice. *Biomass & Bioenergy* 28, 237–248.

Joyce, P.M. (1998) *Growing Broadleaves*. COFORD, National Council for Forest Research & Development, Ireland.

Junfeng, Li and Runqing, Hu (2003) Sustainable biomass production for energy in China. *Biomass & Bioenergy* 25, 483–499.

Jurvélius, M. (2004) Forest fires (prediction, prevention, preparedness and suppression). In: Burley, J., Evans, J. and Youngquist, J. (eds) *Encyclopedia of Forest Science*, Vol. 1. Elsevier, Oxford, pp. 334–339.

Kahle, H-P., Karjalainen, T., Schuck., A., Ågren, G.I., Kellomäki, S., mellert, K., Prietzel, J., Rehfuess, K-E. and Spiecker, H. (2008) *Causes and Consequences of Forest Growth Trends in Europe*. European Forest Institute Research Report 21, Brill, Leiden, 261pp.

Kanowski, P.J. (1997) Afforestation and plantation forestry for the 21st century. In: *Productive Functions of Forests. Proceedings of the XI World Forestry Congress*, 13–22 October 1997, Antalya, Turkey. Ministry of Forestry, Ankara, Vol. 3, pp. 23–34.

Kanowski, P.J. and Borralho, N.M.G. (2004) Economic returns from tree breeding. In: Burley, J., Evans, J. and Youngquist, J. (eds) *Encyclopedia of Forest Science*, Vol. 4. Elsevier, Oxford, pp. 1561–1567.

Karnosky, D.G. and Thakur, R.C. (2004) Genetic aspects of air pollution and climate change. In: Burley, J., Evans, J. and Youngquist, J. (eds) *Encyclopedia of Forest Science*, Vol. 1. Elsevier, Oxford, pp. 223–229.

Keeves, A. (1966) Some evidence of loss of productivity with successive rotations of *Pinus radiata* in the south east of S. Australia. *Australian Forestry* 30, 51–63.

Kenk, G. and Guehne, S. (2001) Management of transformation in central Europe. *Forest Ecology and Management* 151, 107–119.

Kerr, G. and Evans, J. (1993) *Growing Broadleaves for Timber*. Forestry Commission Handbook 9. HMSO, London.

Khan, M.A.W. and Chaudhary, N.R. (1961) Eucalypt trials in India. *Proceedings of the 10th All-India Silvicultural Conference*, Dehra Dun, pp. 535–545.

King, K.F.S. (1975) It's time to make paper in the tropics. *Unasylva* 27 (109), 2–5.

Kirschbaum, M.U.F. (2003) Can trees buy time? An assessment of the role of vegetation sinks as part of the global carbon cycle. *Climatic Change* 58, 47–71.

Kjaer, E.D. (2004) Tropical hardwoods breeding and genetic resources. In: Burley, J., Evans, J. and Youngquist, J. (eds) *Encyclopedia of Forest Science*. Elsevier, Oxford, pp. 1527–1532.

Knowles, O.H. and Parrotta, J.A. (1995) Amazonian forest restoration: an innovative system for native species selection based on phonological data and field performance studies. *Commonwealth Forestry Review* 74, 230–243.

Koch, N.E. and Skovsgaard, J.P. (1999) Sustainable management of planted forests: some comparisons between Central Europe and the United States. *New Forests* 17, 11–22.

Koh, M.P. and Hoi, W.K. (2003) Sustainable biomass production for energy in Malaysia. *Biomass & Bioenergy* 25, 517–529.

Kohlin, G. and Amacher, G.S. (2005) Welfare implications of community forest plantations in developing countries: the Orissa social forestry project. *American Journal of Agricultural Economics* 87, 855–869.

Kramer, P.J. and Koslowski, T.T. (1979) *Physiology of Woody Plants*. Academic Press, New York.

Joyce, P.M. (1998) *Growing Broadleaves: Silvicultural Guidelines for Ash, Sycamore, Wild Cherry, Beech and Oak in Ireland.* COFORD, Dublin.

Lam, D.C. (1998) Fixation des dunes vives par *Casuarina equisetifolia* au Vietnam. *Bois et Forệts des Tropiques* 256, 35–41.

Lamb, A.F.A. (1973) *Fast Growing Timber Trees in the Lowland Tropics* No. 6. Vol. 1: *Pinus caribaea*. Department of Forestry, Oxford.

Lamb, D. (1994) Reforestation of degraded tropical forest lands in the Asia-Pacific region. *Journal of Tropical Forest Science* 7, 1–7.

Lamb, D. (1998) Large-scale ecological restoration of degraded tropical forest lands: the potential role of timber plantations. *Restoration Ecology* 6, 271–279.

Lamb, D. (2001) Reforestation. In: Levin, S.A. (ed.) *Encyclopedia of Biodiversity*. Academic Press, San Diego, pp. 97–108.

Lamb, D. and Tomlinson, M. (1994) Forest rehabilitation in the Asia-Pacific region: past lessons and present uncertainties. *Journal of Tropical Forest Science* 7, 157–170.

Lasco, R.D. (2002) Forest carbon budgets in Southeast Asia following harvesting and land cover change. *Science in China Series C – Life Sciences* 45, 55–64.

Lavender, D.P. (1990) Plantation Forestry in Western Canada. *Proceedings of the XIX IUFRO World Congress*, Montreal, Canada, Vol. 1, pp. 23–31.

Lee, S.J. (1999) Predicted genetic gains from Sitka spruce production programmes. *Forestry Commission Information Note FCIN26*. Forestry Commission, Edinburgh.

Lehman, J. (2007) A handful of carbon. *Nature* 447, 143–144.

Leslie, A.D. (2004) The impacts and mechanics of certification. *International Forestry Review* 6, 30–39.

Lewandowski, I., Weger, J., van Hooijdonk, A., Havlickova, K., van Dam, J. and Faaij, A. (2006) The potential biomass for energy production in the Czech Republic. *Biomass & Bioenergy* 30, 405–421.

Leyendecker-Hilders (1910) The restoration of beech woods with conifers. *Allgemeine Forst und Jadzeitung* 86, 384–386.

Liu, D. (ed.) (2003) *Rehabilitation of Degraded Forests to Improve Livelihoods of Poor Farmers in South China*. Center for International Forestry Research, Bogor, Indonesia.

Lonsdale, D. and Gibbs, J.N. (2002) Effects of climate change on fungal diseases of trees. In: Broadmeadow, M. (ed.) *Climate Change: Impacts on UK Forests. Forestry Commission Bulletin 125*. Forestry Commission, Edinburgh, pp. 83–97.

Lowenthal, D. (2000) *George Perkins Marsh – Prophet of Conservation*. University of Washington Press, Seattle.

Lugo, A.E., Parrotta, J.A. and Brown, S. (1993) Loss of species caused by tropical deforestation and their recovery through management. *Ambio* 22, 106–109.

Lund, G.H. (2000) *Definitions of Forest, Deforestation, Afforestation and Reforestation*. Forest Information Services, Manassas, VA.

Maginnis, S. and Jackson, W. (2003) The role of planted forests in forest landscape restoration. Paper presented at UNFF Intersessional Experts Meeting 'Role of Planted Forests in Sustainable Forest Management', 24–27 March 2003, Wellington, New Zealand.

Mahmood, K., Morris, J., Collopy, J. and Slavich, P. (2000) Groundwater uptake and sustainability of farm plantations on saline sites in the Punjab province of Pakistan. *Agricultural Water Management* 48, 1–20.

Maraseni, T.N., Cockfield, G. and Apan, A. (2007) A comparison of greenhouse gas emissions from inputs into farm enterprises in Southeast Queensland, Australia. *Journal of Environmental Science and Health Part A – Toxic/Hazardous Substances & Environmental Engineering* 42, 11–18.

Marcar, N. and Khanna, P.K. (1997) Reforestation of salt-affected and acid soils. In: Nambiar, Sadanandan, E.K. and Brown, A.G. (eds) *Management of Soil, Nutrients and Water in Tropical Plantation Forests*. ACIAR Monograph No. 43. Australian Centre for International Agricultural Research, Canberra, pp. 481–525.

Marchadier, H. and Sigaud, P. (2005) Poplars in biotechnology research. *Unasylva* 56 (221), 38–39.

Martinez-Garza, C. and Howe, H.F. (2003) Restoring tropical diversity: beating the time tax on species loss. *Journal of Applied Ecology* 40, 423–429.

Mather, A. (ed.) (1993a) *Afforestation: Policies, Planning and Progress*. Belhaven Press, London.

Mather, A. (1993b) Review. In: Mather, A. (ed.) *Afforestation: Policies, Planning and Progress*. Belhaven Press, London, pp. 207–219.

Mayers, J. and Bass, S. (2004) *Policy that Works for Forests and People: Real Prospects for Governance and Livelihoods*. Earthscan, London.

McNabb, K. (2004) An introduction to plantation forestry in Brazil. In: Gonçalves, J.L. de M., Benedetti, V. and McNabb, K. (eds) *Forest Nutrition and Fertilization*. Institute of Forest Research and Study (IPEF) Piracicaba, Sao Paulo, pp. 1–12.

Mendham, D.S., O'Connell, A.M. and Grove, T.S. (2003) Change in soil carbon after land clearing or afforestation in highly weathered lateritic and sandy soils of southwestern Australia. *Agriculture Ecosystems & Environment* 95, 143–156.

Midgley, S.J., Turnbull, J.W. and Johnston, R.D. (eds) (1983) *Casuarina Ecology, Management and Utilization*. CSIRO, Melbourne.

Miller, H.G. (1995) The influence of stand development on nutrient demand, growth and allocation. *Plant and Soil* 168–169, 225–232.

Miller, H.G. (2004) Nutrient limitations and fertilization. In: Burley, J., Evans, J. and Youngquist, J. (eds) *Encyclopedia of Forest Science*. Elsevier, Oxford, pp. 1235–1241.

Ministry of Environment (1990) *Developing India's Wastelands*. Ministry of Environment and Forests, New Delhi.

Moffat, A.J. (2004) Reclamation of Mining Lands. In: Burley, J., Evans, J. and Youngquist, J. (eds) *Encyclopedia of Forest Science*. Elsevier, Oxford, pp. 1078–1085.

Moffat, A.J. and McNeill, J. (1994) Reclaiming disturbed land for forestry. *Forestry Commission Bulletin 110*. HMSO, London.

Moffat, A.J., Jones, W.M. and Mason, W. (2006) Managing brash on conifer clearfell site. *Forestry Commission Practice Note FCPN013*. Forestry Commission, Edinburgh.

Moore, K.D. (2005) (Over) stories: examining the philosophical assumptions behind intensively managed forest plantations. *Journal of Forestry* 103, 97–98.

Mori, T., Ohta, S., Ishida, A., Toma, T. and Oka, T. (2000) Overview of the changing forest ecosystems in East Kalimantan. In: Guhardja, E. , Fatawi, M., Sutisna, M., Mori, T. and Ohta, S. (eds) *Rainforest Ecosystems of East Kalimantan. Ecological Studies* 140, 309–317. Springer, Tokyo.

Morris, A.R. (1993) Forest floor accumulation under *Pinus patula* in the Usutu forest, Swaziland. *Commonwealth Forestry Review* 72, 114–117.

Morris, A.R. (2003) Site and stand age effects on fertiliser response in *Pinus patula* plantations in Swaziland. *Southern Africa Forestry Journal* 199, 27–40.

Morris, J. (1997) Water use by eucalypt plantations. *ACIAR Forestry Newsletter* 22, 1 and 4.

Morris, L.A. (2004a) Soil organic matter forms and function. In: Burley, J., Evans, J. and Youngquist, J. (eds) *Encyclopedia of Forest Science*. Elsevier, Oxford, pp. 1201–1207.

Morris, L.A. (2004b) Nutrient cycling. In: Burley, J., Evans, J. and Youngquist, J. (eds) *Encyclopedia of Forest Science*. Elsevier, Oxford, pp. 1227–1235.

Muhammed, N., Koike, M., Sajjaduzzaman, M. and Sophanarith, K. (2005) Reckoning social forestry in Bangladesh: policy and plan versus implementation. *Forestry* 78, 373–383.

Nambiar, Sadanandan, E.K. (1996) Sustained productivity of forests is a continuing challenge to soil science. *Soil Science Society of America Journal* 60, 1629–1642.

Nambiar, Sadanandan, E.K. (1998) Productivity and sustainability of plantation forests. *Proceedings of the Silvotecna X IUFRO conference 'Site productivity improvement'*, Concepcion, Chile, June 1998.

Nambiar, Sadanandan, E.K. and Brown, A.G. (eds) (1997) *Management of Soil, Nutrients and Water in Tropical Plantation Forests*. ACIAR (Australian Centre for International Agricultural Research) Monograph No. 43.

Nawir, A.A. and Santoso, L. (2005) Mutually beneficial company–community partnerships in plantation development: emerging lessons from Indonesia. *International Forestry Review* 7, 177–192.

Neeff, T., Fehse, J., Ebeling, J. and Henders, S. (2007) Evolving markets for the environmental benefits of forestry. *Quarterly Journal of Forestry* 101, 135–144.

Nehra, N.S., Becwar, M.R., Rottmann, W.H., Pearson, L., Chowdhury, K., Chang, S.J., Wilde, H.D. (2005) Forest biotechnology: innovative methods, emerging opportunities. *In Vitro Cellular & Developmental Biology – Plant* 41, 701–717.

Neilson, D. (2007a) Corporate Private Sector Dimensions in Planted Forest Investments. *FAO Planted Forests and Trees Working Paper Series FP/40E*. Food and Agriculture Organization, Rome.

Neilson, D. (2007b) *The Global Tree Farm and Managed Forest Industry Review*. 2007 edition (RISI) (http://www.dana.co.nz).

Nepstad, D.C., Uhi, C. and Serrão, A.S. (1991) Recuperation of a degraded Amazonian landscape: forest recovery and agricultural restoration. *Ambio* 20, 248–255.

Niesten, E., Frumhoff, P.C., Manion, M. and Hardner, J.J. (2002) Designing a carbon market that protects forests in developing countries. *Philosophical Transactions of the Royal Society of London Series A – Mathematical Physical and Engineering Sciences* 360, 1875–1888.

Nilsson, K., Randrup, T.B. and Wandall, B.M. (2001a) Trees in the urban environment: some practical considerations. In: Evans, J. (ed.) *The Forests Handbook. Vol. 2 Applying Forest Science for Sustainable Management*. Blackwell Science, Oxford, pp. 260–271.

Nilsson, K., Randrup, T.B. and Wandall, B.M. (2001b) Trees in the urban environment. In: Evans, J. (ed.) *The Forests Handbook. Vol. 1 An Overview of Forest Science*. Blackwell Science, Oxford, pp. 347–361.

Nisbet, T. (2005) Water use by trees. *Forestry Commission Information Note FCIN065*. Forestry Commission, Edinburgh.

Noble, I. *et al.* (2000) Implication of different definitions and generic issues. In: Watson, R.T., Noble, I.R., Bolin, R., Ravindranath, N.H., Ravado, D.J. and Dokken, D.J. (eds) *Land Use, Land-Use Change and Forestry*. Intergovernmental Panel on Climate Change Special Report. Cambridge University Press, Cambridge, pp. 53–126.

Nonhebel, S. (2005) Renewable energy and food supply: Will there be enough land? *Renewable & Sustainable Energy Reviews* 9, 191–201.

Nussbaum, R. (2002) *Group Certification for Forests: A Practical Guide*. Proforest, Oxford (http://www.proforest.net).

O'Loughlin, E. and Nambiar, E.K.S. (2001) Plantations, farm forestry and water. *Water and Salinity Issues in Agroforestry No. 8*. RIRDC Publication No. 01/137. Rural Industries Research and Development Corporation, Canberra.

Olschewski, R. and Benitez, P.C. (2005) Secondary forests as temporary carbon sinks? The economic impact of accounting methods on reforestation projects in the tropics. *Ecological Economics* 55, 380–394.

Ortloff, W. (1999) Sustainability issues in Switzerland's forests. *New Forests* 18, 59–73.

Pandey, D. (2000) Teak: a global overview. *Unasylva* 51 (201): 3–13.

Parish, F., Sirin, A., Charman, D., Joosten, H., Minayeva, T. and Silvius, M. (eds) (2007) *Assessment on Peatlands, Biodiversity and Climate Change: Executive Summary*. Global Environment Centre, Kuala Lumpur and Wetlands International, Wageningen.

Park, Y.S. (2002) Implementation of conifer somatic embryogenesis in clonal forestry: technical requirements and deployment considerations. *Annals of Forest Science* 59, 651–656.

Parris, K.M. and Lindenmayer, D.B. (2004) Evidence that creation of a *Pinus radiata* plantation in south-eastern Australia has reduced habitat for frogs. *Acta Oecologica – International Journal of Ecology* 25, 93–101.

Patzek, T.W. and Pimentel, D. (2005) Thermodynamics of energy production from biomass. *Critical Reviews in Plant Sciences* 24, 327–364.

Pennington, T.D. (1997) *The Genus Inga*. Royal Botanic Gardens, Kew.

Perera, K.K.C.K., Rathnasiri, P.G. and Sugathapala, A.G.T. (2003) Sustainable biomass production for energy in Sri Lanka. *Biomass & Bioenergy* 25, 541–556.

Perera, W.R.R. (1962) The development of forest plantations in Ceylon since the seventeenth century. *Ceylon Forester* 5: 142–147.

Perum Perhutani (1992) Teak in Indonesia. In: Wood, H. (ed.) *Teak in Asia. FORSPA Publication 4. Proceedings of the Regional Seminar, Guangshou, China, March 1991*. FAO, Bangkok.

Peterken, G.F. (1996) *Natural Woodland*. Cambridge University Press, Cambridge.

Piotto, D., Montagnini, F., Ugalde, L. and Kanninen, M. (2003) Performance of forest plantations in small and medium-sized farms in the Atlantic lowlands of Costa Rica. *Forest Ecology and Management* 175, 195–204.

Pizarro, R., Araya, S., Jordan, C., Farias, C., Flores, J.P. and Bro, P.B. (2006) The effects of changes in vegetative cover on river flows in the Purapel river basin of central Chile. *Journal of Hydrology* 327, 249–257.

Poulsen, J. (2001) Designing sustainable large-scale industrial plantation landscapes: use of natural forest remnants and corridors to maintain biodiversity, retain environmental functions/services, and reduce social tensions between plantation companies and local people. Paper presented at Proceedings of the International Symposium on Rehabilitation of Degraded Forest Environment for the Next Generation, Department of Forest Resources, Sangju National University, Sangju, Korea, 23–27 April 2001.

Poulsen, J. (2002) *Typology of Planted Forests*. CIFOR Infobrief, CIFOR, Bogor.

Powers, R.F. (2001) Assessing potential sustainable wood yield. In: Evans, J. (ed.) *The Forests Handbook. Vol. 2 Applying Forest Science for Sustainable Management*. Blackwell Science, Oxford, pp. 105–128.

Prestemon, J.P. and Abt, R.C. (2002) The Southern timber market to 2040. *Journal of Forestry* 100, 16–22.

Puri, G.S. (1960) *Indian Forest Ecology*, Vols 1 and 2. Oxford Book and Stationery, India.

Purnomo, H. and Guizol, P. (2003) Governing forest plantation to reduce poverty and improve forest landscape: a multiagent simulation approach. In *Modsim 2003: International Congress on Modelling and Simulation*, Townsville, Australia. Vol. 3: *Socio-Economic Systems*, pp. 1054–1059.

Ramachandra, T.V., Kamakshi, G. and Shruthi, B.V. (2004) Bioresource status in Karnataka. *Renewable & Sustainable Energy Reviews* 8, 1–47.

Ravindranath, N.H. and Hall, D.O. (1994) Indian forest conservation and tropical deforestation. *Ambio* 23, 521–523.

Raymond, D.H. and Wooff, W.G. (2006) Small-scale forest plantations are the key to the future of the Solomon Islands forest industry. *International Forestry Review* 8, 222–228.

Reed, F.L.C. (1983) Forest renewal in Canada. *Commonwealth Forestry Review* 62, 169–177.

Rejmánek, S.H. and Richardson, D.M. (2003) Invasiveness of conifers: extent and possible mechanisms. In: Mill, R.R. (ed.) *Proceedings of the Fourth International Conifer Conference*, August 1999, Wye, UK. *Acta Horticulturae* 615, 375–380.

Rennie, P.J. (1955) The uptake of nutrients by mature forest growth. *Plant and Soil* 7, 49–55.

Richards, K.R., Sampson, R.N. and Brown, S. (2006) *Agricultural and Forestlands: US Carbon Policy Strategies*. Pew Center on Global Climate Change, Arlington, VA.

Rishi, P. (2002) Leading JFM through nurturance: and Indian scenario. *International Forestry Review* 4, 143–148.

Roberts, J. and Rosier, P. (2007) The water resource implications of broadleaved afforestation. *Quarterly Journal of Forestry* 101, 21–27.

Rodwell, J. and Paterson, G. (1994) Creating new native woodland. *Forestry Commission Bulletin 112*. HMSO, London.

Ryan, T. and Shea, G. (1977) *Exotic pine plantations in Queensland and the role of the Beerwah-Beerburrum forests*. Post Consultation Tour Notes No. 8. Third World Consultation of Forest Tree Breeding, Canberra.

Saigal, S. (2005) Joint management of state forest lands: experiences from India. In: Garforth, M. and Mayers, J. (eds) *Plantations, Privatisation, Poverty and Power – Changing Ownership and Management of State Forests*. Earthscan, London, pp. 175–199.

Sajjakulnukit, B. and Verapong, P. (2003) Sustainable biomass production for energy in Thailand. *Biomass & Bioenergy* 25, 557–570.

Salleh Mohd, N. (2000) High value timber species: prospects and challenges. *JIRCAS Working Report* 16, 7–16.

Santos, T., Telleria, J.L., Diaz, M. and Carbonell, R. (2006) Evaluating the benefits of CAP reforms: can afforestations restore bird diversity in Mediterranean Spain? *Basic and Applied Ecology* 7, 483–495.

Sargent, C. and Bass, S. (eds) (1992) *Plantation Politics: Forest Plantations in Development*. Earthscan, London.

Savill, P.S. and Evans, J. (2004) Thinning. In: Burley, J., Evans, J. and Youngquist, J. (eds) *Encyclopedia of Forest Science*, Vol. 2. Elsevier, Oxford, pp. 845–850.

Savill, P.S., Evans, J., Auclair, D. and Falck, J. (1997) *Plantation Silviculture in Europe*. Oxford University Press, Oxford.

Saxena, N.C. (2001) The new forest policy and joint forest management in India. In: Evans, J. (ed.) *The Forests Handbook*, Vol. 2. Blackwell Science, Oxford, pp. 233–259. (http://www.consecol.org/vol5/iss2/art14/. Accessed July 2009)

Saxena, K.G., Rao, K.S., Sen, K.K., Maikhuri, R.K. and Semwal, R.L. (2002) Integrated natural resource management: approaches and lessons from the Himalaya. *Conservation Ecology* 5, 14.

Sayer, J. (2002) Restoration: the challenge for 21st century foresters. *Bois et Forêts des Tropiques* 272, 3–4.

Scotcher, J.S.B. (2006) Forest certification in South Africa. *Southern Africa Forestry Journal* 206, 1–3.

Scott, D.F., Bruijnzeel, L.A., Vertessey, R.A. and Calder, I.R. (2004) Impacts of forest plantations on streamflow. In: Burley, J., Evans, J. and Youngquist, J. (eds) *Encyclopedia of Forest Science*. Elsevier, Oxford, pp. 367–377.

Sedjo, R.A. (2004) International trade in wood products. In: Burley, J., Evans, J. and Youngquist, J. (eds) *Encyclopedia of Forest Science*. Elsevier, Oxford, pp. 1857–1863.

Shan, J.P., Morris, L.A. and Hendrick, R.L. (2001) The effects of management on soil and plant carbon sequestration in slash pine plantations. *Journal of Applied Ecology* 38, 932–941.

Simpson, J. (1900) *The New Forestry – or the Continental system Adapted to British Woodlands and Game Preservation*. Pawson and Brailsford, Sheffield.

Simula, M. (2002) Fibre supply: the Achilles heel of the global pulp and paper industry. *Paperi Ja Puu – Paper and Timber* 84, 16–19.

Sinclair, F. (2004) Agroforestry. In: Burley, J., Evans, J. and Youngquist, J. (eds) *Encyclopedia of Forest Science*. Elsevier, Oxford, pp. 27–32.

Skovsgaard, J.P. and Henriksen, H.A. (1996) Increasing site productivity during consecutive generations of naturally regenerated and planted beech (*Fagus sylvatica* L.) in Denmark. In: Spiecker, H., Meilikaainen, K., Kohl, M. and Skovsgaard, J.P. (eds)

Growth Trends in European Forests. European Forest Institute Research Report No. 5. Springer, Berlin, pp. 89–97.

Smith, C.K., Oliveira, F.D., Gholz, H.L. and Baima, A. (2002) Soil carbon stocks after forest conversion to tree plantations in lowland Amazonia, Brazil. *Forest Ecology and Management* 164, 257–263.

Smith, D.M. (1986) *The Practice of Silviculture.* John Wiley, New York.

Smith, J. and Scherr, S.J. (2003) Capturing the value of forest carbon for local livelihoods. *World Development* 31, 2143–2160.

Song, Z. (1991) A review of the development of shelterbelt system in China. In: Shi, K. (ed.) *Development of Forestry Science and Technology in China.* China Science and Technology Press, Beijing, pp. 64–70.

Speight, M. and Wainhouse, D. (1989) *Ecology and Management of Forest Insects.* Clarendon Press, Oxford.

Spiecker, H., Meilikaainen, K., Kohl, M. and Skovsgaard, J.P. (eds) 1996 Growth trends in European forests. *European Forest Institute Research Report No. 5.* Springer, Berlin.

Stainback, G.A. and Alavalapati, J.R.R. (2005) Effects of carbon markets on the optimal management of slash pine (*Pinus elliottii*) plantations. *Southern Journal of Applied Forestry* 29, 27–32.

Stickel, P.W. (1928) Physical characteristics and silvicultural importance of podsol soil. *Ecology* 9, 176–187.

Stone, M.W. (2006) China and certification: a questionable future. *Journal of Forestry* 104, 332–334.

Strauss, S.H., Campbell, M.M., Pryor, S.N., Coventry, P. and Burley, J. (2001) Plantation certification and genetic engineering: FSC's ban on research is counterproductive. *Journal of Forestry* 99, 4–7.

Strauss, S.Y. (2001) Benefits and risks of biotic exchange between *Eucalyptus* plantations and native Australian forests. *Austral Ecology* 26, 447–457.

Streets, R.J. (1962) *Exotic Forest Trees in the British Commonwealth.* Clarendon Press, Oxford.

Sudha, P., Somashekhar, H.I., Rao, S. and Ravindranath, N.H. (2003) Sustainable biomass production for energy in India. *Biomass & Bioenergy* 25, 501–515.

Sun, G., Zhou, G.Y., Zhang, Z.Q., Wei, X.H., McNulty, S.G. and Vose, J.M. (2006) Potential water yield reduction due to forestation across China. *Journal of Hydrology* 328, 548–558.

Sutton, B.C.S., Attree, S.M., El-Kassaby, Y.A., Grossnickle, S.C. and Polonenko, D.R. (2004) Commercialisation of somatic embryogenesis for plantation forestry. In: Walter, C. and Carson, M. (eds) *Plantation Forest Biotechnology for the 21st Century.* Research Signpost, Kerala, pp. 275–301.

Thorne, J. (1989) Solutions to soil erosion. *New Scientist* (3 June), 45–49.

Thornley, J.H.M. (2007) Modelling forest ecosystem: the Edinburgh forest model. In: Reynolds, K.M., Thomson, A.J., Kohl, M., Shannon, M.A., Ray, D. and Rennolls, K. (eds) *Sustainable Forestry: from Monitoring and Modelling to Knowledge Management and Policy Sciences.* CAB International, Wallingford, pp. 342–373.

Troup, R.S. (1952) *Silvicultural Systems*, 2nd edn. Jones, E.W. (ed.). Clarendon Press, Oxford.

Tucker, N.I.J. and Murphy, T.M. (1997) The effects of ecological rehabilitation on vegetation recruitment: some observations from the wet tropics of North Queensland. *Forest Ecology and Management* 99, 133–152.

Turnbull, J.W. (1983) The use of *Casuarina equisetifolia* for protection forests in China. In: Midgley, S.J., Turnbull, J.W. and Johnston, R.D. (eds) *Casuarina Ecology, Management and Utilization.* CSIRO, Melbourne, pp. 55–57.

Turnbull, J.W. (2002) Tree domestication and the history of plantations. In: Squires, V.R. (ed.) *The Role of Food, Agriculture, Forestry and Fisheries and the Use of Natural Resources*. Encyclopedia of Life Support Systems developed under auspices of UNESCO. Eolss Publishers, Oxford.

Turnbull, J.W. (2007) Development of sustainable forestry plantations in China: a review. *ACIAR Impact Assessment Series Report No. 45*. Australian Centre for International Agricultural Research, Canberra.

Turner, J., Gessel, S.P. and Lambert, M.J. (1999) Sustainable management of native and exotic plantations in Australia. *New Forests* 18, 17–32.

Twedt, D.J. and Wilson, R.R. (2002) Development of oak plantations established for wildlife. *Forest Ecology and Management* 162, 287–298.

Twedt, D.J., Wilson, R.R., Henne-Kerr, J.L. and Grosshuesch, D.A. (2002) Avian response to bottomland hardwood reforestation: the first 10 years. *Restoration Ecology* 10, 645–655.

Tyler, A.L., Macmillan, D.C. and Dutch, J.C. (1996) Models to predict the general yield class of Douglas fir, Japanese larch, and Scots pine on better quality land in Scotland. *Forestry* 69, 13–24.

Valder, P. (1999) *The Garden Plants of China*. Florigeum, Sydney, Australia.

Van der Zel, D.W. (1995) Accomplishments and dynamics of the South African afforestation permit systems. *South African Forestry Journal* 172, 49–57.

van Eijck, J. and Romijn, H. (2008) Prospects for managing Jatropha biofuels in Tanzania: an analysis of strategic niche management. *Energy Policy* 36, 311–325.

van Frankenhuyzen, K. and Beardmore, T. (2004) Current status and environmental impact of transgenic forest trees. *Canadian Journal of Forest Research – Revue Canadienne de Recherche Forestiere* 34, 1163–1180.

van Goor, C.P. (1985) The impact of tree species on soil productivity. *Netherlands Journal of Agricultural Sciences* 33, 133–140.

Varmola, M. (2002) The importance of hardwood plantations in the tropics and subtropics. *The International Forestry Review* 4, 110.

Vergara, P.M. and Simonetti, J.A. (2006) Abundance and movement of understory birds in a maulino forest fragmented by pine plantations. *Biodiversity and Conservation* 15, 3937–3947.

Vertessy, R.A., Zhang, L. and Dawes, W.R. (2003) Plantations, river flows and river salinity. *Australian Forestry* 66, 55–61.

Wagner, R.G., Little, K.M., Richardson, B. and McNabb, K. (2006) The role of vegetation management for enhancing productivity of the world's forests. *Forestry* 79, 57–79.

Wainhouse, D. (2005) *Ecological Methods in Forest Pest Management*. Oxford University Press, Oxford.

Walker, J. and Reuter, D.J. (eds) (1996) *Indicators of Catchment Health: A Technical Perspective*. CSIRO, Melbourne.

Wallace, J.S. (1996) The water balance of mixed tree-crop systems. In: Ong, C.K. and Huxley, P. (eds) *Tree-crop Interactions: A Physiological Approach*. CAB International, Wallingford, pp. 189–233.

Wang, D. (1988) The history of the ornamental plants of China. *Camellia News* 107, 14–16.

Way, M.J. (1966) The natural environment and integrated methods of pest control. *Journal of Applied Ecology* 3 (Suppl.), 29–32.

WBCSD (2007) *Biofuels: Issue Brief – Energy and Climate*. World Business Council for Sustainable Development, Geneva, Switzerland.

Weidemann, E. (1923) *Zuwachsruckgang und Wuchstockungen der Fichte in den mittleren und den unteren Hohenlagen der Saschsichen Staatsforsten*. Tharandt. (Translation no. 302, by CP Blumenthal, USDA 1936).

Weih, M. (2004) Intensive short rotation forestry in boreal climates: present and future perspectives. *Canadian Journal of Forest Research – Revue Canadienne de Recherche Forestiere* 34, 1369–1378.

Weinland, G. (1998) Plantations. In: Appanah, S. and Turnbull, J.M. (eds) *A Review of Dipterocarps – Taxonomy, Ecology and Silviculture.* Centre for International Forestry Research (CIFOR), Bogor, pp. 151–185.

Wenk, G. and Vogel, M. (1996) Height growth investigations of Norway Spruce (*Picea abies* [L.] Karst.) in the Eastern part of Germany during the last Century. In: Spiecker, H., Meilikaainen, K., Kohl, M. and Skovsgaard, J.P. (eds) *1996 Growth trends in European Forests. European Forest Institute Research Report No. 5.* Springer, Berlin, pp. 99–106.

Weyerhaeuser, H., Wen, S. and Kahn, F. (2006) Emerging Forest Associations in Yunnan, China: Implications for Livelihoods and Sustainability. *Small and Medium Forestry Enterprise* 13. IIED, London.

White, T.L. (2004) Breeding theory and genetic testing. In: Burley, J., Evans, J. and Youngquist, J. (eds) *Encyclopedia of Forest Science*, Vol. 4. Elsevier, Oxford, pp. 1551–1561.

Whyte, A.G.D. (1973) Productivity of first and second crops of *Pinus radiata* on the Moutere gravel soils of Nelson. *New Zealand Journal of Forestry* 18, 87–103.

Wiersum, K.F. (1983) Effects of various vegetation layers of an *Acacia auriculiformis* forest plantation on surface erosion in Java, Indonesia. *Proceedings of the International Conference on Soil Erosion and Conservation*, Malama Aina, Hawaii.

Williams, M. (1993) Afforestation: the United States. In: Mather, A. (ed.) *Afforestation: Policies, Planning and Progress*. Belhaven Press, London, pp. 192–206.

Willoughby, I. (2007) Use dye markers to reduce pesticide use. *Forestry Commission Technical Note FCTN016*. Forestry Commission, Edinburgh.

Wilson, P. (2007) The United Kingdom Woodland Assurance Scheme. *Quarterly Journal of Forestry* 101, 111–114.

Wingfield, M. (2004) Disease affecting exotic plantation species. In: Burley, J., Evans, J. and Youngquist, J. (eds) *Encyclopedia of Forest Science*, Vol. 2. Elsevier, Oxford, pp. 816–821.

Winter, R.K. (1974) *The Forest and Man*. Vantage, New York, USA.

Wong, G.Y. and Alavalapati, J.R.R. (2003) The land-use effects of a forest carbon policy in the US. *Forest Policy and Economics* 5, 249–263.

Woods, R.V. (1990) Second rotation decline in *P. radiata* plantations in South Australia has been corrected. *Water, Air and Soil Pollution* 54, 607–619.

Wormald, T.J. (1975) *Pinus patula. Tropical Forestry Paper No. 7*. Commonwealth Forestry Institute, Oxford.

Worrell, R. and Malcolm, D.C. (1990) Productivity of Sitka spruce in Northern Britain. Prediction from site factors. *Forestry* 63, 119–123.

Wunderle, J.M. Jr (1997) The role of animal seed dispersal in accelerating native forest regeneration on degraded tropical lands. *Forest Ecology and Management* 99, 223–235.

WWF (2002) *Forest Landscape Restoration*. WWF Position Paper, January 2002. WWF International, Gland.

Yang, X. (2001) Impacts and effectiveness of logging bans in natural forests: People's Republic of China. In: Durst, P.B., Waggener, T.R., Enters, T. and Tan, L.C. (eds) *Forests Out of Bounds: Impacts and Effectiveness of Logging Bans in Natural Forests in Asia-Pacific.* RAP Publication 2001/08. Food and Agriculture Organization of the United Nations, Regional Office for Asia and the Pacific, Bangkok, pp. 81–102.

Ying, J.H. and Ying, J.H. (1997) Comparative study on growth and soil properties under different successive rotations of Chinese fir. *Journal of Jiangsu Forestry Science and Technology* 24, 31–34.

Youngs, R.L. (2004) History and overview of wood use. In: Burley, J., Evans, J. and Youngquist, J. (eds) *Encyclopedia of Forest Science.* Elsevier, Oxford, pp. 1852–1856.

Yusoff, M.K., Heng, S.S., Majid, N.M., Mokhtaruddin, A.M., Hanum, I.F., Alias, M.A. and Kobayashi, S. (2001) Effects of different land use patterns on stream quality water in Pasoh, Negri Sembilan, Malaysia. In: Kobayashi, S., Turnbull, J.W., Toma, T., Mori, T. and Majid, N.M.N.A. (eds) *Rehabilitation of Degraded Forest Ecosystems in the Tropics.* Center for International Forestry Research, Bogor, pp. 87–97.

Zerbe, J.I. (2004) Energy from wood. In: Burley, J., Evans, J. and Youngquist, J. (eds) *Encyclopedia of Forest Science.* Elsevier, Oxford, pp. 601–607.

Zhang, Y.X. and Song, C.H. (2006) Impacts of afforestation, deforestation, and reforestation on forest cover in China from 1949 to 2003. *Journal of Forestry* 104, 383–387.

Zobel, B.J., van Wyk, G. and Stahl, P. (1988) *Growing Exotic Forests.* John Wiley and Sons, New York.

Zurita, G.A., Rey, N., Varela, D.M., Villagra, M. and Bellocq, M.I. (2006) Conversion of the Atlantic Forest into native and exotic tree plantations: effects on bird communities from the local and regional perspectives. *Forest Ecology and Management* 235, 164–173.

Index

Note: page numbers in *italics* refer to figures, tables and boxes, those with suffix 'n' refer to Notes.